THE SOCIETY OF GENES

THE SOCIETY OF GENES

Itai Yanai and Martin Lercher

Harvard University Press

Cambridge, Massachusetts
London, England
2016

Copyright © 2016 by Itai Yanai and Martin Lercher
All rights reserved
Printed in the United States of America

First printing

Library of Congress Cataloging-in-Publication Data
is available from the Library of Congress.

ISBN 978-0-674-42502-6

To Michal and Vero, and
to our children

Contents

Preface

*It is not from the benevolence of the butcher,
the brewer, or the baker, that we expect our dinner,
but from their regard to their own interest.*

—Adam Smith

THERE IS AN ANCIENT SOCIETY OF GENES THAT IS INEXTRICABLY linked to our human society. The members of this society shaped your body and your brain, your instincts and your desires. They have brought humanity to the present, but they don't necessarily dictate our future. To understand how these genes influence us—and how humanity can rise above them—you might imagine that we need to find out what each individual gene does.

But this approach won't work, because we are not the simple sum of our genes. The members of the society of genes do not live in isolation. Working together, forming rivalries and partnerships, is the only way they can form a human body that can sustain them for a few decades and propel them into the next generation of humanity.

Almost 250 years ago, Adam Smith realized that it is the self-interested interactions of individuals that make the marketplace efficient. Similarly, it is the competition and cooperation among genes striving for their own long-term survival that promotes the persistence of humanity as a whole.

The previously unimaginable genomic information that modern technology continues to amass has revealed much about the structure of the society of genes. There are the hard-working individuals on the factory floor, such as hemoglobin, which carries oxygen to the furnaces of the cells, and polymerase, which produces faithful copies of other genes. There are messengers, such as the *FGFR3* gene, which registers growth signals and passes them on, and causes bursts of genetic diseases when broken. There are managers, such as *FOXP2*, which commands a workforce involved in human language, and *SOX9*, which, when broken, lets a girl develop in a body that would otherwise have been male. There are huge armies of freeloaders that exploit the other members of the society of genes, such as the LINE1 elements, which litter our genome with half a million copies. And there are downright dangerous characters like some versions of the *BRCA1* gene, which bring their female carriers one step closer to breast cancer.

If we want to understand our genome, the key is an understanding of the strategies of these genes. A genome—we will find—is best seen as a conglomerate of selfish genes, held together by an intricate network of cooperation. This is the story of the society of genes, of the triumphs and failures of its members, and their eternal conflicts and partnerships.

THE SOCIETY OF GENES

Prologue

NEARLY TWENTY YEARS AGO, LONG BEFORE WE MET AT THE
European Molecular Biology Laboratory in Heidelberg, each
of us read Richard Dawkins's 1976 classic *The Selfish Gene.*
That book changed our lives. At the time, we were a computer
scientist and a physicist, but we left those fields to become
evolutionary biologists. Dawkins's book describes a grand
perspective on what living organisms really are: survival machines, "robot vehicles blindly programmed to preserve the
selfish molecules known as genes." This astonishing truth,
hidden from human view by the evolutionary timescale on
which it works, never ceases to amaze us. It may be as difficult to get used to as the weird statistical world of quantum
mechanics, so strange to us because of the tiny length scales
on which it manifests itself.

When Dawkins wrote *The Selfish Gene*, not a single
genome was available for analysis. He had synthesized the
book's logic from basic principles and the work of previous scientists who themselves had built their theories on basic principles. *The Selfish Gene* still stands essentially correct, even

after the genomic revolution. That revolution deposited piles of genome sequences in public databases, giving us a treasure trove of biological information. The first genomic sequences spelled out precisely what the set of genes underlying a survival machine looks like. As the genomes of more and more species were published, comparisons among them provided amazing insight into their similarities and differences. Those insights, in turn, allowed us to deduce how genes evolve. For our own species, the genome sequences of many hundreds of individuals are now available.

As time went on, it became clear that a holistic perspective was necessary if we were to more deeply understand biological systems and their evolution. Genes do indeed behave in ways that can be described as selfish. But genes, like humans, do not live in isolation. No gene can make it by itself. To survive throughout the ages, genes had to cooperate in building and operating one survival machine after the other. All human genomes contain the same genes. But individual copies of a gene may differ due to mutations, and there is fierce competition among competing copies striving for supremacy in the genomes of future generations. Because of their complex interactions, their cooperation and competition, genes are best considered as members of a society, as we argue throughout this book. The selfish gene concept carried us a long way into the third millennium; the next distance will be far more easily covered if we extend this view by considering complete societies of genes. Dawkins clearly had in mind the importance of this perspective. Indeed, a chapter on this topic in Matt Ridley's wonderful book *The Origins of Virtue*, published in 1996, is titled "The Society of Genes" to indicate that survival machines are the product of the coordinated actions of many genes. Yet the interaction among genes was not well-studied enough at that time to allow for a coherent understanding. Marvin Minsky's book *The Society of Mind* provides a theory on how intelligence results from the action of individual agents. Likewise, in this book we show how a genome results from the sum of the relationships among individual genes. In elaborating on the concept of a society of genes, we provide a sweeping overview of biology, starting from the evolution of single cells inside our bodies, and zooming out in space and time until we reach the beginning of life itself.

We wrote this book for a general audience, assuming no background in biology on the part of the reader. But we also hope that it will be interesting to our colleagues by offering a new angle on the evolution of genes and genomes. We would be thrilled if this book would inspire students to direct their curiosity toward the study of genomes, just as we ourselves were inspired by *The Selfish Gene*.

A person dear to us has a special habit of skipping to the end of a novel before he finishes it. If he dies before he has read the entire book, at least he'll know how it ends. While that particular justification seems disproportionate, here we remove any drama by providing a summary of the book, laying down the broad strokes of our argument for the usefulness of the society of genes analogy when thinking about living systems.

We begin with a catastrophic failure of cooperation. Cancer is a disease of the genome, the 6-billion-letter-long "encyclopedia" that contains all the information needed to build us. In discussing how cancer arises, Chapter 1 introduces the major players of the book—the cells that multiply to build your body, the genes and their interactions that control them, and the mutations that change the genes' letter sequences and thereby provide the substrate for evolution. Before cancerous cells become life threatening, they must accumulate several specific mutations that cooperate in accelerating growth, each overriding one of the body's defenses against uncontrolled cell proliferation. It is extremely unlikely that any single cell would pick up these mutations all at once. Why then is cancer so prevalent? The key to this agonizing riddle is the logic of natural selection, first fully described by Charles Darwin. Once a cell obtains just one of the required mutations, it starts to divide faster than its neighboring cells. Eventually, its progeny will be so numerous that the next mutation in one of them becomes likely. In this way, the defenses of the genome against cancer fall like dominoes.

As illustrated by the dynamics of cancer, the genome is not fixed; it changes even over the course of a lifetime. Chapter 2 introduces the analogy of the society of genes—the "community" formed by the various distinct copies of genes found in human genomes. Any society needs to define its borders. The immune systems of bacteria and vertebrates offer two alternative solutions to the problem of distinguishing society members from the genes of potentially dangerous intruders. Both immune systems are based on comparing potentially foreign genes or their products to templates stored in the genome. In contrast to our own immune system, which exploits the principle of natural selection, a bacterium's ingenious system directly introduces information about a current intruder into its own genome—a rare example of the environment directly shaping the genome. On short time scales, such Lamarckian principles also operate in humans: as a mother suckles her baby, she transfers crucial immune protection hard-won from her own experiences. For a gene, a spot in the next generation is worth killing for. That is exactly how "poison / antidote" gene pairs survive—by killing off competing sperm or egg cells that do not carry their copies. Chapter 3 explains the strategy that the society of genes evolved to suppress such cheating as much as possible, giving each gene copy the same chance to pass to the next generation. This thoroughly egalitarian design is necessary to make sex an efficient strategy of propagation. At first glance, sex can seem like a silly idea: instead of cloning themselves, mothers settle with contributing just half of their genome to their offspring, while the other half is provided by a father who may give little else. But on the million-year timescales on which the history of genes unfolds, sex turns out to be a brilliant idea. In a world that is constantly changing, the advantages of trying out new combinations of gene versions in every generation outweigh the costs. The lion's share of genomic novelties arises in fathers—copying mistakes, mostly harmful, that arise during the many rounds of cell duplication in the production of sperm.

Natural selection is an important, but not the only, agent influencing the fate of gene copies in the society. Simple chance plays just as big a role, as Chapter 4 describes. Consider this apparent contradiction: the genomes of any two humans on this planet are roughly 99.9 percent identical, and yet humans often treat other humans almost as if they belonged to another species. The tiny differences we find between the genomes of people from different regions tell us how humans spread from Africa to the rest of the world over the course of the past 100,000 years, but they also tell tales of adaptations to specific regions. Skin color—a fine balance between ultraviolet (UV) radiation protection and the use of sunlight to make vitamin D—and lactose tolerance—related to dairy farming—are two examples of how the environment determines which version of the same gene is beneficial enough to become established in a region. But most genomic differences have no practical relevance whatsoever—the corresponding gene copies are neutral bystanders, often hitchhiking on the evolutionary success of the society members in their genomic neighborhood. Natural selection promotes genes that cause you to discriminate against those who are, in genetic terms, relatively distant from you, the very idea of racism. But make no mistake: the genes responsible for this behavior are promoting their own selfish interests, even if this runs against what's best for you or for humanity as a whole.

The genes of the society form a complicated web of relationships. Performing a given task usually requires the cooperation of several genes, and most genes have several different responsibilities. Chapter 5 describes the complexity of the organization chart connecting our genes to our features. While many human genetic diseases can be pinpointed to the malfunctioning of a single gene, more typically, diseases follow from distorted interactions among multiple members of the society of genes, often in combination with the environment. Moreover, due to multifunctionality, different mutations to the same gene may lead to symptoms as drastically different as a sex reversal and a facial deformity. Complicated interactions extend far beyond the origin of diseases and govern the societies of genes of organisms as different as humans and bacteria. The society of genes cannot stand still. When a new river splits a society into two by preventing local populations on opposite shores to mix, the two cannot help but separate over time into distinct species. At the heart of the formation of new species, we find two societies of genes whose members can no longer cooperate. In Chapter 6, we frame the evolution that led to modern humans and chimpanzees as such a splitting of an ancestral society. Where was the breaking point preventing proto-humans and proto-chimps from mixing their societies of genes? While speculation on human-chimp hybrids ("chumans") is now the subject of tabloids, a close comparison of human and chimp genomes indeed reveals ancient scandals. Analogous scandals between new-species-to-be also occurred more recently: millions of years after the last chuman, "modern" humans reencountered Neanderthals outside Africa. Although we usually depict the Neanderthals indigenous to Europe and Asia as apelike brutes, there must have been substantial attraction between them and the new arrivals. Intimate encounters left traces in our genomes that still help us fight the germs of Europe and Asia.

The members of the society of genes can be broadly classified into managers and workers. Much innovation can be implemented by managing the same genes slightly differently. Accordingly, Chapter 7 argues that different species are distinguished more by their management than by their worker genes. Changes in management contributed to such innovations as human speech or our larger brains. The HOX genes are the highest-level construction managers of your body and that of most other animals. When broken by mutations, a HOX gene can put legs on a fly's head. Bacteria have no legs or heads, but some of them can morph into resilient time capsules to survive by lying dormant during tough times. The manager gene that oversees this transition is none other than a distant cousin of the HOX genes.

How does the society recruit new genes? Chapter 8 describes how the duplication of genes allows for their diversification into new functions. As a consequence, much of our genome consists of modified copies of other genes. This duplication paradigm has led to spectacular successes, such as color vision, based on three copies, and our sense of smell, based on hundreds. Bacteria make frequent use of a related strategy to expand their society of genes—they copy genes from the genomes of other bacteria, a form of intellectual theft. This allows them to pick up new society members to defend themselves against antibiotics or to quickly tap new food sources.

Societies can split to form new species. But they can also fuse, with spectacular consequences. Chapter 9 shows how our cells are the product of a billion-year-old merger between two very different types of bacteria. In this symbiotic relationship, the merged society of genes was able to evolve in directions inaccessible to either of its parents. Each of your cells is in principle nothing but a blown-up bacterium of one type (an *archaebacterium*), housing lots of bacteria of the other type *(eubacteria)* to supply its energy, with the tenants' genomes well mixed into the landlord's due to billion-year-long intense contact. As in successful corporate mergers, a fusion can be much stronger than the sum of its parts.

Freeloaders are an unavoidable threat to any society. As Chapter 10 shows, for the past 4 billion years, a spectacular diversity of freeloaders has survived by exploiting cellular life forms. One result of freeloading off the society of genes is the exaggerated size of the human genome. It is burdened by an exuberance of genes able to copy/paste themselves within the genome, able to stay in the society of genes without contributing to human survival. Such freeloading is a general phenomenon: the genome of the onion is five times as big as yours because of such parasitic sequences. The ancestors of viruses—the mothers of all freeloaders—must have already mingled among the first simple organic molecules, which may have assembled 4 billion years ago in rock cavities around deep-sea vents.

This provides only a rough sketch of what follows. To savor the details, you'll need to read on.

Evolving Cancer in Eight Easy Steps

With great power comes great responsibility.
—Voltaire

BOB MARLEY AND THE WAILERS TURNED THE WORLD ON TO reggae and inspired millions to think spiritually about their ways of life. Tragically, Marley's career was cut short at the age of thirty-six, when his body succumbed to skin cancer. The cancer began seemingly harmlessly underneath a toenail four years earlier, and Marley attributed it to a soccer injury. When his doctors insisted that the toe had to be amputated, he wouldn't hear of it, citing his Rastafarian interpretation of the Old Testament verse: "They shall not . . . make any cuttings in the flesh." Unchecked, the tumor spread in a relentless progression, driven by the simple principle of natural selection among the cells of the body. Had Marley understood the way that cancer evolves, he might have had the tumor removed in time—he might have lived to attend his induction into the Rock and Roll Hall of Fame in 1994.

Cancer may be the scariest of all major diseases; it certainly is the one most difficult to prevent and treat. While modern medicine can overcome many other diseases through drugs, such pharmaceutical strategies are fraught with considerable difficulties for cancer. What makes the agent that causes cancer so difficult to target?

Cancer is not an attack on the body from the outside, nor is it simply a terrible accident that occurs within the body. Instead, it is a manifestation of the power of evolution. It follows an inescapable logic, identical to that which governs the evolution of animal and plant species. As a prelude to the story of the society of genes, this chapter introduces cells, genes, and evolution through the lens of this dangerous disease.

Cancerous tumors are an integral part of our body, which makes preventing and treating them so difficult. We can view a human body as a building made from trillions of building blocks called cells. Cells exchange nutrients and chemical signals. Each cell is akin to a tiny factory, with different types of cells performing specialized functions, all of which contribute to the working of the entire body. In cancer patients, some of the cells abandon their cooperation with the rest of the body and start to multiply uncontrollably.

The cells that make up your body form a pedigree. New cells arise when existing cells divide into two. You can place all your body's cells on one gigantic family tree, rooted at the single cell that began your existence: your mother's fertilized egg. Figure 1.1 illustrates what the cell lineage of a worm, a much simpler animal than you, looks like. This tree displays the development of an animal through cell divisions, starting from a single cell.

The single fertilized egg cell develops into a full human being is achieved without a construction manager or architect. The responsibility for the necessary close coordination is shared among the cells as they come into being. It is as though each brick, wire, and pipe in a building knows the entire structure and consults with the neighboring bricks to decide where to place itself.

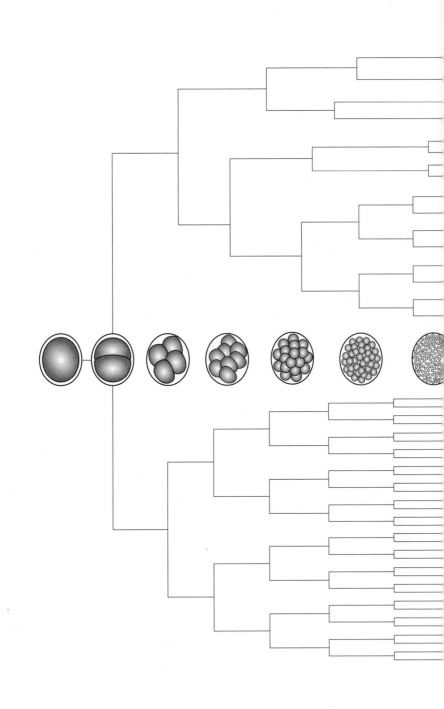

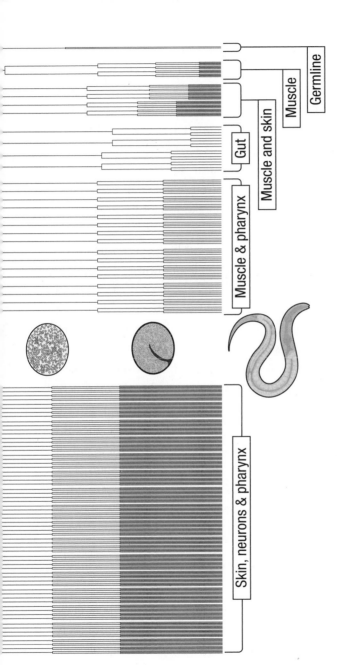

Figure 1.1: The cell lineage of a tiny roundworm called *Caenorhabditis elegans*. This simple animal develops in just thirteen hours from a single cell (top) to an animal with 558 cells (bottom). At the center are drawings of the embryo as it develops, with circles indicating cells. The family tree of cells that builds the worm—its cell lineage—flanks the drawings. Each vertical line indicates a growing cell, while a horizontal line is a cell division. Groups of cells specialize in building particular organs, such as the throat (pharynx) or nerve cells (neurons). Human cell lineages are much larger and more complex but follow the same principles. We discuss the germline in Chapter 3.

Cancer is one branch in the person's cell lineage, an overgrown mass of the patient's own cells. Every cancer starts out as a single cell on that lineage. The cell and its successors keep dividing past the point they should normally have stopped. The multiplying cancer cells spread through the body, securing their access to vital resources, such as oxygen. Eventually, they become so prevalent and consume so much of the body's resources that other parts of the body can collapse from starvation, sealing a catastrophic breakdown of the division of labor among our many cells.

How do the cells that make up your body know when to divide and when to stop? Cell division is an intricate process that requires fine-tuning. If half of the cells in your face or hand divided just once more, you would vaguely resemble Joseph Merrick, who made his living in the nineteenth century by exhibiting himself as the Elephant Man. Such unwanted cell divisions are usually kept in check by a local democracy: cells only grow and divide when signaled to do so by other cells surrounding them. The cells communicate through growth factors, specific messenger molecules produced inside the cells and then sent out through the cell walls. A cell divides only if it receives simultaneous signals from different neighbors. This integration of multiple signals is a safeguard, protecting the body from misjudgments made by individual cells.

A Disease of the Genome

Cancer cells multiply without regard for their neighbors' cues, because they are different from other cells. At the core of every cell—of anything alive—is the genome, a set of vulnerable molecules called chromosomes. Each human genome can be viewed as a text consisting of 6 billion letters, a thousand times larger than the complete works of Shakespeare. These letters are split into forty-six volumes, each volume being one chromosome. Our genome appears to come with its own backup copy. The forty-six chromosomes are twenty-three sets of almost identical chromosome pairs, with the only exception being the two unpaired sex chromosomes of males, named X and Y. The genomic text is written in an alphabet of just four letters: A, T, C, and G, shorthand for four molecules called nucleobases (or simply bases): A for adenine, T for thymine, C for cytosine, and G for guanine. Millions of bases are linked into a chain, resulting in a type of molecule called deoxyribonucleic acid, or DNA (Figure 1.2).

Chromosomes are composed of two DNA molecule strands running alongside each other and tightly interlinked (Figure 1.2). The strands are complementary mirror images of each other: every A on one strand is faced by a T on the other, and every C is faced by a G. To display genomic information as a stretch of letters, we need look at only one of the two strands, since we can easily reconstruct the other strand by applying the complementary mirror rules. To get an idea of what your genome's text looks like, consider this tiny fragment of human chromosome 9:

. . . ACCAGTTCTCCATGATGTGAATTTTCCA
TTGTATGACTGAACCACAATATCTCAGGG
ACCCCATAAATAT . . .

This string of letters is not in itself very informative, and we are still far from knowing how to interpret every letter in this text. As of now, no single genome has been completely

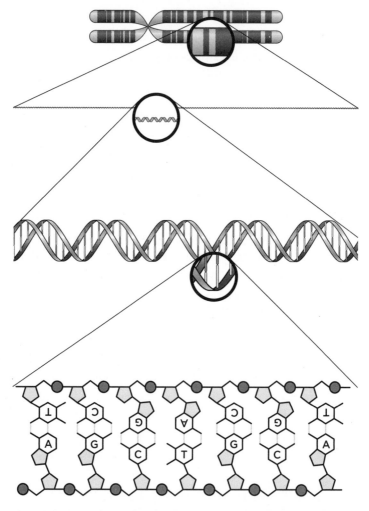

Figure 1.2: A chromosome is a giant molecule composed of millions of copies of the four DNA bases (or letters), A, T, C, and G, connected into two complementary chains (or strands). Throughout the life cycle of a cell, chromosomes change their shape. The top panel shows them in their most condensed form, where the DNA strands are tightly packed. The lower panels show a chromosome section at increasing magnification. The bottom panel shows the two complementary strands, where A always pairs with T, and C with G.

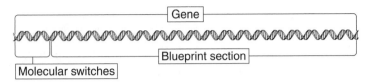

Figure 1.3: A protein-coding gene consists of the blueprint used as a template for making the protein (the coding sequence), as well as the molecular switches used to turn on and off the production of the protein.

deciphered; we simply do not yet fully understand all that it encodes.

Texts in any human language can be divided into paragraphs representing more or less coherent thoughts. Similarly, we can partition any genome into discrete fragments of coherent information, called genes. About 20,000 of the human genes contain precise, blueprint-like instructions on how to make large molecules called proteins, which perform most of the specific functions carried out in a cell. These protein-coding genes are the protagonists of this book.

The letter sequences of these genes can be classified into two categories (Figure 1.3): a blueprint section (the coding sequence), which describes the makeup of the protein, and a section of molecular switches. These switches modulate the activity of the gene, controlling whether and at what rate the blueprint section is copied into templates for making proteins. Most genes possess several such switches. As we go along, keep in mind that a gene includes both the instructions for building a protein and the switches that control under which conditions that gene is turned on and off.

As our cells divide to produce more cells, one of the most important things each cell has to do is to make a copy of each of its chromosomes. To make a copy of the double-stranded chromosomal DNA, a cellular machinery called DNA polymerase, composed of dozens of individual proteins, pulls the two strands apart and assembles a new complementary mirror image onto each of the resulting templates.

A chromosome can be damaged or broken, just like any other molecule. A simple example is iron: when oxygen atoms invade and connect to the atoms in pure iron, they form rust molecules. It is easy to distinguish rust from healthy iron, and rust converters can be used to chemically transform rust into a protective chemical barrier. Many types of change to a chromosome are similar to rust in that they can easily be detected and repaired. But there are other modifications to the DNA, where one letter is accidentally replaced with another letter. Such mutations are often overlooked by the proteins responsible for error checking and remain unrepaired. Here is an example of a mutation in the genome sequence we saw before:

> (before the mutation): . . . ACCAG_T_TCTCCATGAT GTGAATTTT . . .
> (after the mutation): . . . ACCAG_C_TCTCCATGAT GTGAATTTT . . .

The sixth letter in this sequence of bases has changed from T to C. Such a small change may seem insignificant. But consider what a single typo can do by looking at this well-known example:

> (before the mutation): . . . MY KINGDOM FOR A HO_R_SE . . .

(after the mutation): . . . MY KINGDOM FOR A
HO_USE . . .

A full 1 percent of our genes are involved in proofreading
and correcting chromosomes. But despite this substantial investment in error correction, DNA copying is not perfect. In
each cell division, there are 6 billion letter pairs to copy, check,
and correct. The chance for one specific DNA letter pair to
be mutated in one such duplication round (the mutation
rate) is about 1 in 10 billion. Thus, in each round of genome
copying, there is still about a 70 percent chance that at least
one pair of letters will have a typo. This number is optimistic,
because it assumes that you lead a healthy life. Your genomes
can suffer even more changes through toxic chemicals (such
as those in cigarette smoke or burned meat) or exposure to ultraviolet radiation (from the sun or visits to tanning salons).
Such mutations can be single letters exchanged for others, as
in the example above, but in some cases entire segments of the
DNA molecule (whole stretches of letters) are removed or are
duplicated and added at seemingly random places. Because of
these mutations, you do not have one single genome, but instead many billions of slightly different genomes, one for each
of your cells.

With each round of genome copying, errors accumulate.
This is analogous to alterations in medieval books, which were
copied by hand. Each time a copy was made, some changes
were introduced inadvertently; over time, as changes accumulated, the copies may have accrued meanings at variance with
the original. Similarly, genomes that have undergone more
copying processes will have accumulated more mistakes. To

make things worse, mutations may damage those genes responsible for the proofreading and repair of genomes, further accelerating the introduction of mutations.

Most mutations do not have any noticeable effects, just like changing the *i* for a *y* in "kingdom" would not distort the word's legibility or meaning. But sometimes a mutation to a human gene results in, for example, an eye whose iris is of two different colors. Similarly, almost everyone has birthmarks, which are due to mutations that occurred as our body's cells multiplied to form skin.

But if mutations are changes to the genome of one particular cell, how can a patch of cells in an iris or a whole patch of skin, consisting of many individual cells, be affected simultaneously? Were all of the millions of cells in the iris of a girl with a blue patch in an otherwise brown eye struck by the same mutation? The answer lies in the cell lineage: if the mutation occurred early on in the lineage of the developing iris, then all cells in that patch have inherited that change (Figure 1.4).

All the cells in such a patch can trace their pedigree back to the same single ancestral cell. Every cell in the patch inherited the same color-changing mutation from that ancestor, and so these cells jointly show a new color that distinguishes them from their nonmutated neighbors. There was one cell in the developing iris of a girl with a two-colored eye that was struck by a color-changing mutation. Likewise, there was a mutation in one cell in the developing blood vessels of somebody with a port-wine stain (a *nevus flammeus*) that caused the abnormal dilation of these vessels, coloring the surrounding skin dark red.

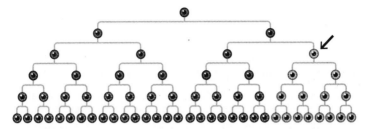

Figure 1.4: A lineage of cells that compose an iris. A mutation that destroys a gene encoding a pigment occurs (arrow) and is inherited by all descendants in this part of the iris, leading to a sector of lighter color in the iris.

Much of our genome copying occurred *in utero,* but many cells are continually renewed throughout life. Skin cells, for example, are replaced by fresh ones on a monthly cycle. Mutations that upset the balance of skin pigments become increasingly likely as we age, explaining the emergence of age spots.

The genomes of cancer cells contain mutations that have much more serious consequences than birthmarks or eyes of different colors. An important cancer-related mutation was identified in an experiment on mouse cells. Under special circumstances, cells can be modified to survive outside the organism, suspended in a fluid of nutrients that contains growth factors (the signals for the cell to grow, which are normally excreted by neighboring cells). Such cells are called cell lines and are invaluable for studying how cells work. Researchers working with mouse cell lines found that they can become cancerous from just one specific mutation: a G replacing a T at a particular position along a gene called *H-Ras* was enough to make them grow even in the absence of growth factors. The discovery was monumental: it showed that misguided instructions from one mutated gene are enough to turn normal cells into cancerous ones.

H-Ras normally functions as part of a response system that reacts to growth factors excreted from neighboring cells. The protein encoded by the *H-Ras* gene acts as a molecular switch: if activated through a chemical modification, it activates other proteins that transmit the signal for growth throughout the cell. The protein encoded by *H-Ras* is normally active only when the cell receives growth factors signaling it to divide. An *H-Ras* mutation can render the protein permanently in the active mode, meaning that the cell divides continually, independent of signals from neighboring cells. *H-Ras* is a normal gene with an important function, but a single mutation can turn it into a cancer gene, what is called an oncogene.

A limit on cell division independent of growth factors is hardwired into the genome, but the limit can be broken. Cell lines are able to divide indefinitely. They bypass the cell division limit, one of the reasons a single mutation to the *H-Ras* gene was enough to make a mouse cell line cancerous. Let's take a look at how this happens.

Duplication counters are attached to both ends of each chromosome, enabling the cell to keep a rough track of how many cell divisions it has undergone along its family tree. The ends of each chromosome, called telomeres (from the Greek word for "end"), are made up of a specific sequence of letters: TTAGGG, repeating itself several thousand times. Just imagine:

. . . TTAGGGTTAGGGTTAGGGTTAGGGT
TAGGGTTAGGGTTAGGGTTAGGGTTAGG

GTTAGGGTTAGGGTTAGGGTTAGGGTTAG
GGTTAGGGTTAGGGTTAGGGTTAGGGTTAGGGT
TAGGGTTAGGGTTAGGGTTAGGGTTAGGGTTAGG
GTTAGGGTTAGGGTTAGGGTTAGGGTTAG
GGTTAGGGTTAGGGTTAGGGTTAGGGT
TAGGGTTAGGGTTAGGGTTAGGGTTAGG
GTTAGGGTTAGGGTTAGGGTTAGGGTTAG
GGTTAGGGTTAGGGTTAGGGTTAGGGT
TAGGGTTAGGG . . .

When a chromosome is copied, these telomeres get shorter. When the same chromosome is copied again, the telomeres get shorter still. The decreasing length of the telomeres is inherent in the way chromosomes are copied: in making each copy, a section at the end of the telomere is left behind. It is as if each chromosome was handed a ticket for multiple duplication trips, with one section of the ticket torn off every time it enters a new cell. After several dozen rounds of copying, the ticket is used up—the telomere is worn down completely—and the chromosome can no longer be replicated.

A human cell without telomeres is programmed to eventually commit suicide. This is a good thing: worn-down telomeres are a sign of uncontrolled proliferation, and cell suicide acts as a fail-safe switch to protect the rest of the body. A cancer cell needs to avert this suicidal program; it must find a way to rebuild its telomeres. Its solution is simple: it enlists the help of a complex molecular machine called telomerase, which specializes in rebuilding telomeres (Figure 1.5). Telomerase is assembled from several proteins (or subunits), encoded in separate genes across several chromosomes.

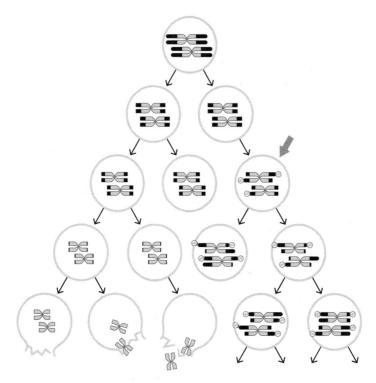

Figure 1.5: The circles and arrows indicate a lineage of cells. The telomeres (black segments) at the ends of chromosomes shorten with each cell division until worn down; at this point the cell commits suicide (left lineage). However, if a mutation occurs (arrow) that turns the telomerase gene on, the telomeres are rebuilt, and the cell lineage can continue to proliferate (right lineage).

Given that telomere shortening is a mechanism against un-controlled cell division, it seems odd, on the face of it, that the genome would contain a gene set that overrides this safe-guard. Looking more closely, we see that this override is nec-essary. For example, to ensure that the next human generation starts with full-length telomeres, the telomerase must rebuild the chromosome end sections lost in making egg and sperm cells. The telomerase is therefore under lock and key: it is for use only in a privileged subgroup of "immortal" cells, such as those occupied with the production of sperm and egg cells and is disabled for other cells that might turn into cancers.

But recall that every cell has the same genome. The genes for telomerase are already present as passive bystanders in every cell. All that is needed for cancer to progress beyond the point of telomere loss is a mutation at the right position at the beginning of the *TERT* gene, the gene responsible for the crucial subunit of telomerase. Part of the letters that constitute the *TERT* gene specify the building instructions for the telomerase subunit. These instructions are not changed by the mutation that activates telomerase. Instead, the mutation modifies a group of letters that constitute a molecular switch dedicated to the gene's regulation. While these letters usually specify that telomerase is only to be turned on in particular cells, such as the sperm precursors, a mutation can change the switch so that telomerase is produced in the cancer cell. Telomerase is activated in roughly 90 percent of cancers; the remaining cancers use an alternative way to stabilize their telomeres.

Telomeres protect the ends of the chromosomes, which otherwise would stick together. By the time a given cell mutates to switch on its telomerase, the telomeres are already worn down and the chromosomes are clumping up. This is one of the reasons why when you look at cancer cells under a microscope, you see that they contain irregular chromosomes (Figure 1.6).

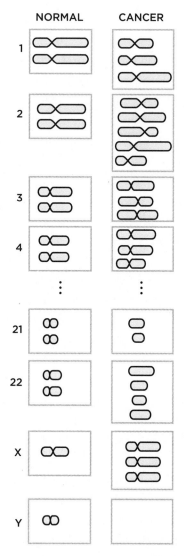

Figure 1.6: A comparison between the set of chromosomes of a normal cell (left) and a cancerous cell (right). The disorganized state of the cancer cell chromosomes is mostly due to the instability following telomere shortening before the telomerase gene is turned on.

Cancer's Wish List

Mutations to *H-Ras* and the *TERT* telomerase gene are but two examples among many mutations that contribute to the emergence of cancer. The exact mutations differ widely across cancer types and patients, but their effects can be classified into a set of recurring properties. These properties are the hallmarks of cancer, as they have been famously called by cancer researchers Douglas Hanahan and Robert Weinberg. Each hallmark describes one way in which mutations override the body's safeguards against uncontrolled cell growth:

1. *Self-supplying growth signals.* A cell in the adult body divides only when signaled to do so by growth factors secreted from its neighboring cells, a process akin to peer pressure. Becoming cancerous requires a mutiny: the cell needs to supply itself with orders to replicate. The *H-Ras* mutation belongs to this general class.

2. *Ignoring anti-growth signals.* Cells also receive signals from their neighbors to stop dividing. A cancer cell needs to turn a deaf ear to these signals.

3. *Becoming immortal.* The genome limits the number of consecutive cell divisions through telomere shortening, and cancerous cells need to override this mechanism. *TERT* mutations account for the majority of mutations in this class.

4. *Evading cell suicide.* The cell is engineered to realize when things have gone horribly wrong, at which point it sets off a chain of events culminating in its own

destruction. A cancer cell needs to avert suicide, so it has to remove this mechanism.

5. *Evading immune destruction.* One of the tasks of the immune system is to find and destroy cancerous cells before they can spread. If a tumor is to survive, it must escape detection by the immune system.

6. *Consuming energy greedily.* Uncontrolled cell proliferation must be fueled accordingly. Cancer cells switch to a mode that extracts energy from sugars more quickly, but also more wastefully, increasing the burden on the rest of the body.

7. *Attracting new blood vessels.* Cells use the bloodstream to receive vital oxygen. If a cell continues to divide without securing its daughter cells' supply of oxygen, then the newly divided cells will starve. A cancerous cell needs to induce nearby blood vessels to grow toward it.

8. *Invading distant body parts.* A cancer is most dangerous when it finds a way to travel from its place of origin, infiltrating new body parts to set up way stations.

These hallmarks accumulate gradually. Only full-blown cancers display them all. But how do these hallmarks accumulate? That a single *H-ras* mutation was able to cause cancer in the mouse cells seems at odds with the list of eight hallmarks and also at odds with the slow process of cancer, which often takes decades to emerge.

Cancer is chiefly a disease of the elderly. A seventy-year-old person is more than ten times more likely to develop a malignant tumor than a seventeen-year-old. A good example of the slowness of cancer development is the link between the rise of cigarette smoking and lung cancer. Across the United States, the use of cigarettes doubled each year in the 1920s. The incidence of lung cancer followed with an almost parallel increase—but with a delay of almost thirty years.

If just a single mutation to H-Ras was enough to cause cancer in the mouse cell line, why then do human cancers typically develop over many years, progressing in several stages? The answer to this apparent mystery lies in the nature of the mouse cells used in the original experiments. They were not normal cells; cells used in laboratories rarely are. Certain tricks are required to "immortalize" cells so that they continue to grow in a dish, and these tricks invariably include specific changes to the genome, such as circumventing the telomere shortening. As became clear only later, the mouse cells used in the original experiments were already on the edge of becoming cancerous: just a single step—one mutation to the H-Ras gene—away. For normal human cells to develop into a full-blown cancer, mutations establishing all eight hallmarks need to occur in succession.

One Renegade Genome

The root of cancer lies in the fact that each cell has much more information than is strictly necessary for it to perform its functions. The power bequeathed by this information to each cell can lead to catastrophic failures. If the information is just slightly distorted through mutations, the cell divides when it should not, and if its daughters and granddaughters keep dividing, the body's balance is tipped.

All cancers start out small, beginning with one single, misguided cell, "one renegade cell," to quote Robert Weinberg. But the real renegade is the genome. Any individual cell has a short life cycle of little overall importance. It is the cell's genes that transcend the cell's lifespan. The molecules that constitute the cell will fall apart with time, but the genes may live on. The genes' essence is information, transmitted from cell generation to cell generation. In each of your body's cells, the fate of all the genes on the genome is tightly linked. The genes rise and fall together, with their success crucially dependent on cooperation among them. The genes of a cancerous genome are led astray by a small mutated minority among their number, breaking the laws of controlled cell division to secure themselves an unfair advantage.

There is no single mutation that converts a genome to that of a full-fledged tumor. Cancer, like almost all other processes in your body, requires the genome's genes to act as a team. Each of the eight steps toward cancer overrides an independent defense of the organism. But consider the chances for a given genome to accumulate all eight mutations. For example, what is the probability of picking up a mutation that would circumvent the safeguards against activating the telomerase? If we assume that there are about ten different mutations that could do the job, and with a probability of one in 10 billion for each base to mutate in one cell division, the chance that one mutation occurs that topples one safeguard is approximately one in a billion. So to get all eight mutations in the same genome copying process is a chance of one in a billion, multiplied eight times with itself, which comes to a one-in-a-billion-billion-billion-billion-billion-billion-billion-billions chance. This is as unlikely as hitting the Mega Millions lottery jackpot nine times in a row, which you can safely assume will not happen to you.

Even though these odds are very low, people do get cancer. How do all of the defenses of the organism fall? The answer is that the renegade genome changes slowly, one step at a time.

It is exceedingly unlikely for an individual genome to simultaneously acquire all the mutations necessary to become a full-fledged cancer, but it is not difficult to imagine a single genome in your body getting a mutation that overrides just *one* of the body's defenses. Since your body contains trillions of similar (but not identical) genomes, many different mutations are already present. On average, every second cell division introduces a new mutation into the newly formed genome. Given all of these mutations, it is nearly inevitable that at some point, in one individual genome, a mutation will occur in the wrong place, and the cell will progress one step toward cancer.

Each such mutation changes the game. The crucial idea to grasp is that although a genome needs to unite all the cancer hallmarks to cause a full-fledged cancer, the acquisition of each individual hallmark already makes an important difference. A renegade genome with just one of these mutations may already divide faster due to that mutation. For example, it may outgrow its sister genomes, because the mutation reduces its reliance on growth factors excreted by the surrounding cells. Once the cell starts dividing without its neighbors' approval, it can spawn millions of clone genomes. And it is this change in numbers that enables the next step in the progression. Now that millions of similar renegade genomes are present, the chance increases that *one* of them will acquire the next mutation toward cancer (Figure 1.7). The millions-strong population of renegade genomes may produce one new genome that has two of the eight mutations required for a full-fledged cancer. When that happens, a second defense has fallen.

The genome with two of the hallmark mutations will be even better at propagating itself. The cells carrying this double-mutant genome will divide even faster than their single-mutant sister cells. When the descendants of the double mutant are again millions of cells strong, the chances for the next mutation will again increase. The process continues until the entire repertoire of the organism's defenses has been subverted.

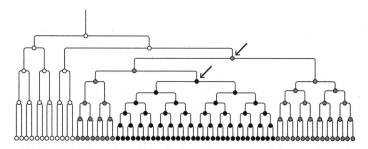

Figure 1.7: How cancer evolves, one step at a time. Time runs from the top (at the first ancestral cell of this cell lineage) to the bottom (all the descendants currently alive). Each arrow indicates a mutation that leads to a faster rate of division relative to the other cells (seen in the faster succession of generations along the branch of the tree below the mutation). With time, the descendants of this faster-dividing cell become abundant enough for the occurrence of a new mutation to become likely. The new mutation will further accelerate cell proliferation. The cycle repeats until one cell has completed all of the steps required for a full-fledged cancer.

This is why cancers take so long to establish themselves, and why many cancers could be averted if only we were able to notice their precursors—the double, triple, or quadruple mutants—in time. Many renegade genomes multiply deep in the body and stay unnoticed until it is too late. But sometimes the renegade cells are visible from the outside, as they were underneath Bob Marley's toenail. Understanding the significance of this could have saved Bob Marley from a premature death. Removing the skin cancer cells would have been the only way to stop the tumor from progressing further. Similarly, while a birthmark is usually harmless, half of all melanomas arise from preexisting birthmarks. A birthmark that increases in size may indicate that it already acquired some of the mutations necessary to turn it into a cancer. When this happens, doctors often recommend the removal of the birthmark as a preemptive strike.

The step-by-step progression of cancer exploits none other than the law of natural selection, the law that mediates the adaptation of all forms of life. Natural selection changes flowers to make them more attractive to bees or hummingbirds. Because of natural selection, bacteria evolve resistance to antibiotics. Natural selection causes moths to take on a different color to blend into their changing environment. Charles Darwin was the first to fully recognize how life changes from generation to generation according to this law. His experiences as a naturalist led him to the insight that all living organisms are related, like the leaves on a great tree's many tangled branches. Darwin's greatest achievement is his clear description of natural selection's beautifully simple logic. It is no exaggeration to say that this process is what spawned all the wonders of the living world.

Darwin arrived at the theory of natural selection by studying plants and animals. Today we can see the same theory in action by observing cells and their genomes. If Darwin had been able to observe cancer cells through a microscope, he would have had just as much evidence for his theory as he obtained by studying complete organisms. Natural selection is a general law that holds whenever members of a group differ in some heritable way, and when these differences affect their chances of leaving offspring. The individuals must belong to a group, or, in more precise terms, a population, that evolves together. A population can be an entire animal species (that is what Darwin studied), all the cells in a human body (as in the case of cancer), or even simple molecules in a test tube that are capable of replicating themselves.

Darwin's conditions for natural selection acting on a certain attribute (for example, to what degree a cell relies on its neighbors' signals in its decision to divide) are that it (1) varies among the individuals of a population, (2) is heritable, and (3) affects fitness. In the context of evolution, fitness means short-term reproductive success, which is the rate at which offspring are spawned relative to what is normal for the population. If these three conditions are met, then, over time, the proportion of the high-fitness cells (those spawning more descendants than average) will increase. It is a logical necessity that the reproductively more successful (that is, fitter) cells eventually take over the population.

In a population of cells, the majority waits for their neighbors' calls to divide before they spawn daughter cells, but a minority may not. As we have seen, such a difference is genetically encoded and open to change, by, for example, mutations to the *H-Ras* gene. Because cells with such mutations will have more descendants, they will eventually dominate the population of cells around them. Cancer progresses according to the law of natural selection, unless the remaining cells find a way to stop the renegades from dividing further, perhaps by directing their body to a doctor to have the renegades removed.

Cancer cannot progress unless all three conditions for natural selection are met. If every cell were identical, the makeup of the population could not change. If cells differed in their rate of proliferation, but this difference was not heritable, then the fast-dividing cells would not become more numerous over time. If cells differed in a heritable way, but this was not connected to their fitness, then the makeup of the population would not change systematically over time, either.

The competition among cells that is inherent in natural selection is, of course, not advantageous for the organism as a whole, nor is it advantageous for the cancer cells in the long term. The cancer dies with the organism. There is no way for a gene with a newly arisen mutation that contributed to the rise of a cancer to jump into the next human generation. That gene is trapped in its cancer cells and cannot make its way into sperm or egg cells (cancers can occur in the testes, but the multiple genomic changes that drive cancer evolution make it impossible to produce functioning sperm). While there are no exceptions to the short-lived fate of cancer cells in humans, there are rare examples in animals of cancer cells managing to transcend the lifetime of the bodies in which they evolved. A tumor that first evolved in one of the earliest domesticated dogs is still around. Today, the descendants of those tumor cells live on the skin of dogs, transmitted from one dog to another through contact—in essence, the tumor has turned into a parasitic species.

In biology, success is measured in terms of long-term survival: the successful genes are those that are still around, continuing to propagate copies of themselves. In that sense, if we set aside the extremely rare exceptions of transmissible tumors, cancerous growth is not in the long-term best interest of any gene, and the mutated gene's success is halted abruptly with the demise of the body. However, the logic of natural selection is short-sighted. Since the cells of the body collectively fulfill the requirements for natural selection, it will occur, and the evolution of cancer becomes almost inescapable. As disturbing as it is, if a human body lives long enough, it will inevitably develop cancer.

Natural selection is not the only parallel between the evolution of cancer and the evolution of species. Here is how biologist Jerry Coyne described life's evolution: "Life on Earth evolved gradually beginning with one primitive species— perhaps a self-replicating molecule—that lived more than 3.5 billion years ago; it then branched out over time, throwing off many new and diverse species; and the mechanism for most (but not all) of evolutionary change is natural selection." This sentence succinctly captures the five principles of evolution: (1) species change, (2) species are related to one another, (3) changes occur gradually, (4) the mechanism for many changes is natural selection, and (5) not all evolutionary change is due to natural selection.

These principles were first sketched out to describe the evolution of species, but they can all equally be applied to the evolution of cancer in an organism's population of cells. The cells of our bodies accumulate changes to their genes from cell generation to cell generation (principle 1: change occurs). Each of us is a colony of cells, all of which descended from a single cell with a single set of genes—the fertilized egg. In cancer, the gene set controlling one renegade cell lineage starts to follow its own agenda, abandoning its cooperation with the rest of the body. This sublineage of cells can be considered a new "species" relative to the noncancerous cells (principle 2: common descent). But no single mutation can transform a perfectly healthy cell into a cancerous one—instead, the renegade genomes accumulate changes slowly, one by one (principle 3: evolution occurs gradually). The proportion of the body taken up by a given cell lineage can change due to heritable mutations—pre-cancerous cells that divide faster outcompete their well-behaved neighbors (principle 4: natural selection). Not all genomic changes are relevant to the function or the proliferation of the cell, so some changes may become common in a population by chance alone (principle 5: chance changes exist).

Speaking of Genes

In this book, we often write about genes as though they had intentions, as though they had consciousness. They don't, of course. Genes are nothing but stretches of DNA, complex assemblies of atoms. But when we examine the genes' properties and their consequences, it appears as if the genes were acting to ensure their own survival. This is because the evolution of genes, like all living things, is driven by the logical necessity of natural selection. For example, when we write "a cancer gene aims to secure an unfair advantage," we are using shorthand for "mutations to an oncogene that cause an increased growth rate of the cells that carry the mutation will over time lead to an increase in the total fraction of body cells that carry the mutation." Anthropomorphizing provides a convenient shorthand for discussing many processes; while this helps to develop an intuition about natural selection, we need to remember the full description behind the shorthand.

One Step Forward, One Step Back

Cancer does not pass from parent to child. The cancer-promoting gene versions evolve in cells of the body other than the sperm and egg cells. The latter have their own, non-cancerous genomes, ensuring that cancer cannot be directly transferred from parents to offspring. But mutations that establish individual steps toward cancer in specific cell types *can* be inherited.

One such example is breast cancer, which is associated with mutations that break two genes, the *breast cancer* genes 1 and 2 (*BRCA1* and *BRCA2*). Women who inherit such a mutation from one of their parents have an 80 percent risk of evolving breast or ovarian cancer over the course of their lives. *BRCA1* and *BRCA2* work together to repair damaged chromosomes, and they initiate a cell's suicide if the chromosomes appear damaged beyond repair. Mutations to these genes apparently lead to functional changes allowing the cells that have them to better avert suicide; one of the hallmarks of cancer is thus already established. It is not yet clear why damaged copies of these genes cause cancer only in breasts and ovaries, but we know that any woman who inherits them starts life with one step toward cancer already completed.

Why then does a cancer genome need to acquire exactly eight hallmarks? Why do there seem to be just enough safeguards of the organism to keep most cancers at bay until we are beyond our forties? It is as if the genome had set up precisely the right number of defenses to postpone the evolution of cancerous tumors until after our reproductive years. This may be what actually happened. The eight safeguards that have to be overcome by evolving cancer cells evolved in our ancestors by natural selection. If our anticancer system were any less efficient, then cancers would claim many more human victims in their twenties and thirties, the age at which humans have produced most of their offspring throughout evolution. Imagine that at some point in the ancient history of our species, there had indeed been fewer safeguards against cancer, but there was one woman with a mutation that provided a better anticancer measure. She would have been able to delay cancer beyond the prime age of reproduction and leave more offspring. The three requirements of natural selection— variability, heritability, and fitness effects—would have been met, and over time, this better cancer defense system would have spread through the whole of humanity.

However, in preindustrial times, before there were antibiotics, few humans survived long enough for full-fledged cancers to evolve, and by the time they died, most of their reproductive work was done. Thus, from the perspective of natural selection, there was no powerful "need" to have a stronger system than the one we possess now. That is, there was no natural selection to promote the evolution of a ninth line of defense in humans.

The case of the naked mole rat makes for an interesting comparison. Small animals typically live for only a few years, but this East African rodent has a life expectancy of thirty years—more than ten times that of its similarly sized cousin, the house mouse. In relative terms, the naked mole rat's longevity is akin to discovering an ape species with a lifespan of 600 years.

Years of observation have not revealed a single cancer in a naked mole rat. In contrast, much cancer research is done on mice, which employ the same eight safeguards against cancer as we do. That naked mole rats reach such biblical ages without evolving cancers likely means that they have either fortified one of the existing safeguards, or that they hit upon a ninth line of defense. Fully understanding the details of the mole rats' solution to the threat of cancer might one day form the basis for new cancer remedies.

Evolution is not something that only happened in the past. It is happening everywhere, all of the time, even within the confines of your own body. Because of that, it is inevitable that we get cancer. But it is not inevitable that we must die of it. Because cancers are so ubiquitous and threatening, cancer research is among the most prominent and thus most advanced fields of research in the life sciences. New therapies appear regularly for specific kinds of cancers. One recent development, immunotherapy, may even prove to be a general breakthrough in treating all kinds of cancer. Here, the body's own defense system is harnessed to fight off the specific lineage of cancer cells arising in the individual. It is conceivable that in the not-so-distant future, cancer might be considered a chronic disease, similar to HIV, which has lost much of its horror for those people with access to advanced healthcare. The immune system, which may hold the key to powerful anticancer therapies, is the topic of Chapter 2. In its context, we also introduce the society of genes analogy as a tool for understanding evolution.

● 2 ●

How Your Enemies Define You

*It is only shallow people who do not judge
by appearances.*

—Oscar Wilde

THE YEAR IS 1993. SIX MIT GRADUATE STUDENTS WALK INTO A
Las Vegas casino with a plan to bring down the house. Sitting
at the blackjack tables, they employ a cheating strategy
called "counting cards," in use since the 1700s. To influence
the odds in their favor, the students need to find out whether
the blackjack table is "hot," that is, how many face cards (kings,
queens, and jacks) have not yet been dealt. Five students
place small bets, each sitting at a different table and counting
the cards. The sixth student stands aside. When one of the
counter's tally shows that a table is hot, they signal to the sixth
student, who then finds a place at that table and begins to
place large bets. By employing this strategy in several casinos,
the students net $3 million.

Casinos do not permit the use of the counting-cards strategy
because of the unfair advantage it gives. Over time, casinos developed more and more sophisticated counterstrategies for
detecting and blacklisting card counters. The simplest strategy
is to stop cheaters who have been caught once from reentering
the casino. For the casino, it is "fool me once, shame on you;

fool me twice, shame on me." In the past, guards were charged with detecting cheaters based on a blacklist; today, large casinos use camera systems coupled to face-recognition computer programs. You can view a casino and its honest customers as a loosely defined society; cheaters try to exploit this society. Keeping the cheats out comes down to a question of establishing borders. To protect itself from exploitation, a society needs to distinguish insiders from outsiders. Immune systems, charged with protecting the body from pathogens, also need to distinguish friend from foe. Long ago, fueled by the power of natural selection, they evolved ways to do so, ways that, strikingly, use the same process that allows cancers to proliferate.

The Society of Genes

In this book, we argue that the genes that make up your genome are best viewed as a society. The human genome contains 20,000 genes, each of which is a specialist in one or more tasks. Genes need to cooperate to build and run a body capable of replicating them. These feats require an intricate organization and fine-tuned division of labor. It would be a mistake, though, to conclude that the genes' coexistence is a sign of genetic harmony.

While every human genome contains essentially the same set of genes, the genes themselves are not identical. As a result of mutations, genes come in different versions, termed alleles. For example, half of the population may have a C in the fourth position of a particular gene, where the rest of humanity has a G. The two alleles distinguished by these letters may impart slightly different functions, such that the bearers of the C allele fare better than their competitors. Over many generations, the G allele may slowly be pushed out of existence.

The gene is analogous to a specialized branch of the economy in a human society (bakeries, pharmacies, or DIY stores, for instance). The different alleles compete with each other, just as different bakeries would compete in a given economy (Figure 2.1). If Betty's Bakery bakes the tastiest crois-

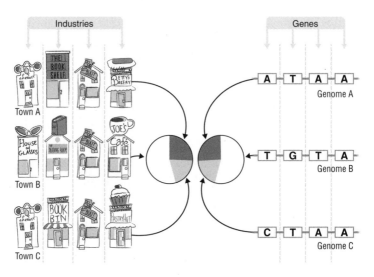

Figure 2.1: The society of genes analogy. On the left are three rows of shops in different shopping malls. Each column of shops represents one type of business in the retail industry: glasses, books, boots, and bakeries. Some stores, such as Bob's Boots, are more successful than others, reflected in their higher number of stores. The left pie chart shows the overall market share of the different bakeries. On the right are three rows of alleles located in different human genomes, depicted as blocks and distinguished from one another by representative mutations. Each column of alleles represents one gene. Some alleles, like the A allele in the rightmost gene, are more successful than others. The right pie chart indicates the relative frequencies of the three alleles of the leftmost gene among all human genomes.

sants in town, their business will likely expand, with some of their competitors going out of business.

The society of genes is the collective sum of all alleles for all genes, occurring anywhere in the genomes of a given population. Your own genome, with the alleles present on its forty-six chromosomes, represents one way of assembling a complete set of instructions for building and operating a body. Alleles can build human bodies in countless other ways, as we know from the myriad differences among us (many of which are inherited through our genes). The current standing of an allele is measured by its popularity in the society: the more human genomes carry a given allele, the more successful we may consider that allele to be.

Just as car manufacturers depend on reliable delivery from their suppliers, each allele's survival hinges on the proper functioning of its peers. Alleles compete in an environment shaped by the rest of the society. For example, two genes may together build a particular molecular machine. Two alleles of these cooperating genes may work especially well together, forming a coalition to propel their mutual success through the survival of the individuals in which they occur. This is reminiscent of profitable arrangements between different businesses, such as between specific coffee and bookstore chains. More generally, we expect to see competition between alleles for the same gene and cooperation between the alleles of different genes. The complex interactions among the society's genes and the insights into life they provide form the central motif of this book.

In thinking about evolution occurring inside our bodies, such as in the case of cancer (Chapter 1) and of the adaptation of our immune system to pathogens (this chapter), we witness short-term evolution. These processes teach us about important functional relationships among the genes, but we do not really see their evolution as a society. This is because, inside our bodies, new cells always arise as clones of existing cells, inheriting identical copies of their mother cells' genomes. Thus, two alleles in different cells will never meet—the cells' societies are static. From the perspective of the genes, the body is of no significance, and that is why we need to look at long-term evolution if we are to learn how the society of genes operates.

The society of genes is where evolution takes place. Individual genomes come and go, but it is the gradual successes and failures of the genes over eons that capture evolutionary changes. What rules govern this society? Alleles are not altruistic idealists. Natural selection will reward an allele by increasing its popularity (its "market share") in the society when the allele enhances the fitness of its carriers. Therefore, each allele "works" toward its own advantage when cooperating with its peers, exemplifying Adam Smith's hypothesis that self-interest, if channeled appropriately, maximizes the common good.

How Bacteria Hold a Grudge

An immune system that breaks down leaves us at the mercy of our enemies. This is what happens in the case of AIDS—acquired immunodeficiency syndrome—and what makes it so dangerous. The HIV virus, which causes AIDS, lives inside human cells whose task it is to protect the body from pathogens, and it manipulates those immune cells to its own advantage. As a consequence, the victim's immune system not only is helpless against HIV, but it also becomes too weak to fend off many threats that healthy people deal with easily, from bacterial and fungal infections to cancers.

All viruses, from HIV to those that cause common colds, are highly accomplished at cheating at the game of self-replication. For a cell to reproduce itself requires a complicated and intricate process, but viruses take a shortcut. Lacking the necessary genes to replicate on their own, viruses survive by freeloading on other societies of genes. A virus enters your body through contaminated food (rotavirus, causing gastroenteritis), through sneezed droplets (rhinovirus, causing common colds), or through the exchange of bodily fluids (HIV, causing AIDS). The virus attaches itself to one of your cells. It then secures its own genome's entrance into the cell's interior and starts hijacking the cell's machinery, redirecting it to make copies of the viral invader. When viral genome copying has exhausted all of the cell's resources, the horde of newly created viruses makes its great escape. You could not accuse them of being sentimental about their hijacked mother ship: many viruses kill the infected cell by bursting it open. Some of the released viruses find new cells to infect, continuing their cycle of propagation through destruction (Figure 2.2).

Bacteria are also besieged by viruses. Consisting of a single cell, bacteria are tiny organisms that have only one genome. The bacterial cell is built much like a human cell, but it is much smaller and simpler in structure. If you imagine that each of your cells is an apartment divided into rooms designed for different functions (kitchen, bedroom, living room), then a bacterium resembles a doghouse. A bacterial genome typically

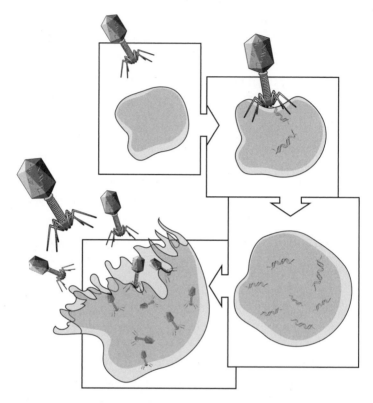

Figure 2.2: The life cycle of a virus. The virus attaches itself to a cell and injects its genome, which then directs the cell's machinery to make copies of the virus. As soon as a large number of viruses has been assembled, the cell bursts apart, and the viruses are released.

contains between 2,000 and 4,000 protein-coding genes, which is five to ten times fewer than ours. The first bacterial genome sequence was deciphered in 1995, and many thousands have been studied since.

Many bacterial genomes incorporate an odd-seeming structure, a region of about thirty DNA letters in a specific sequence, repeated up to one hundred times. These repeat regions account for as much as 1 percent of a bacterium's genome and are nearly palindromic, meaning that they read almost the same forward and backward. These genomic repeats do not sit shoulder to shoulder but are separated by what the discoverers of these structures dismissively called "spacers." Unlike the repeats, the spacer elements vary in length from twenty-five to forty letters long.

For many years, nobody knew what these *repeat-spacer-repeat-spacer-repeat-spacer-repeat* . . . segments were for. But the segments had to have a purpose, since bacteria tend to lose sequences they don't need. Researchers gave the name CRISPRs to these segments for *c*lustered *r*egularly *i*nterspaced *s*hort *p*alindromic *r*epeats. The breakthrough in understanding the function of these anomalous regions came not from understanding what the conspicuous repeats are doing, but rather from a closer look at the seemingly useless spacers. The letter sequences of spacers were often found to be identical to parts of the letter sequences of known viral genomes. But why would a bacterial genome contain bits and pieces of viral information, carefully organized between repeats?

It turns out that these viral fragments are, in effect, mug shots of the bacterium's past aggressors, posted inside each bacterial cell like photos of cheaters are displayed at casinos (Figure 2.3). The bacterium uses this information to recognize and eliminate intruders that resemble previous offenders, effectively immunizing itself from future attacks by close relatives of the viruses it has already encountered. This kind of bacterial immunity against viruses illustrates how a society of

Figure 2.3: Like a guard checking suspects against a collection of mug shots, bacteria compare the genomes of potential intruders to the genomes of past aggressors, whose profiles are stored in the CRISPR spacers.

genes defines its borders: the bacterium maintains a database of what it is *not*, integrating a new mug shot into the genome whenever it detects a previously unknown enemy.

When a researcher infects a colony of unfortunate bacteria with a virus, most of the bacteria will die (Figure 2.4). When the dead bacteria are compared to the genomes of the surviving ones, typically there is only one difference between them: one additional spacer and one additional repeat in the CRISPR region of the survivors, with the new spacer being a perfect complementary mirror image of a segment of the virus. The spacer got there thanks to the work of a gene specializing in integrating bits of viral DNA into the CRISPR structure. Only a few bacteria manage to achieve this on time, explaining why most are still wiped out by the rapidly replicating viruses.

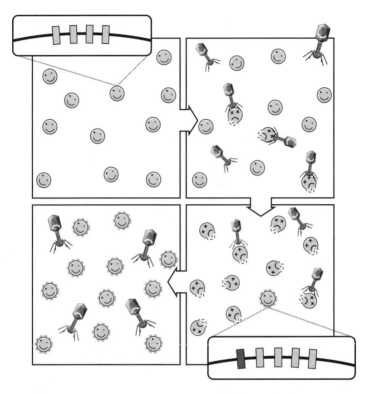

Figure 2.4: The CRISPR region of a bacterium before and after viral infection. A virus attacks the bacterial population, killing off all but one bacterium. The survivor has managed to integrate DNA complementary to a piece of the virus genome into its CRISPR region. This integration enables the bacterium to destroy the viral DNA and thereby immunize itself against the virus. The survivor's offspring inherit the immunity and thrive.

The manner in which bacteria compare mug shots against potential threats exploits the same forces that hold together the two DNA strands of a chromosome. Recall that DNA strands are made of millions of linked letters, the four bases A, T, C, and G. The molecular shapes of our bases are like matching puzzle pieces. Adenine (A) is attracted by chemical forces to thymine (T), and cytosine (C) is chemically attracted to guanine (G). The same pairing rules apply to a very similar type of molecule known as ribonucleic acid, or RNA, except that RNA replaces T with the chemically similar uracil (U). RNA is used to temporarily store information for making templates used in the production of proteins. Some viral genomes are made up of RNA rather than DNA. If you were to put matching single strands of DNA or RNA into a test tube, you would find that when they bump up against each other, they stick to their complementary mirror images, forming double-stranded DNA or RNA.

The CRISPR system exploits this principle of mirror-image attraction. The bacterium copies (or transcribes) the spacers and the flanking repeats of its genome into single-stranded RNA molecules, which it then parades around the cell. Each of these RNA molecules will bind to any matching viral genome that happens to be around (Figure 2.5), and the resulting double-stranded molecule will attract specialized proteins that chop the bound pair to bits.

CRISPR is famous in the life sciences for a reason other than forming this adaptive immune system for bacteria. When the CRISPR system memorizes a new pathogen, it inserts a specific DNA sequence into a specific location in the genome. This functionality has been adapted as an extremely useful tool for research. Using it, you can edit a genome, for example by removing specific genes and observing what happens in their absence.

Had the CRISPR worked perfectly as an immune system for bacteria, there would be no viruses left to threaten bacteria, and the system would have become obsolete. The battle,

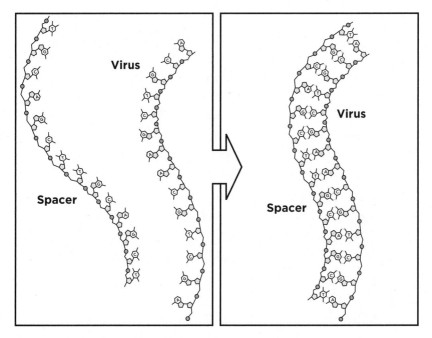

Figure 2.5: The bacterial immune system copies the mug shots stored as CRISPR spacers into single-stranded RNA molecules. These are then attracted to any viral genome sequences that are their complementary mirror images, exploiting the same chemical forces that hold the two strands of our chromosomes together.

however, rages on. Defensive measures by the bacteria lead to countermeasures by the viruses, resulting in an evolutionary arms race.

A virus can evade the bacterial immune system in several ways. The simplest of these ways depends on the fact that, for its surveillance effort to be manageable, the immune system has to lose its oldest spacers as new spacers are added. Viruses that have been absent long enough that the bacteria have "forgotten" them can slip past the bacterial immune systems once again. Another tactic is for the virus to alter its appearance so that it no longer matches its mug shot. All it takes is a single letter mutation in the part of a virus's genome that matches a bacterium's spacer. The bacterium then responds with counter-countermeasures, integrating updated versions of viral mug shots into its genome.

Sometimes a bacterium accidentally integrates a fragment of its own DNA as a CRISPR spacer. Based on this accidental mug shot, the bacterium will mistake its own DNA for that of an aggressor and destroy it in an inadvertent suicide mission—a bacterial autoimmune disease.

Can a bacterial society of genes be faced with more enemies than it can record? There is some evidence that for bacteria living in such areas as the open sea, the genomic real estate required for recording all potential threats would become prohibitively expensive, and CRISPR systems lose their effectiveness.

A Random Mug-Shot Generator

If your body defended itself from intruders with a CRISPR-like task force, a single cell inside your body might gain immunity, but it would have no way of transmitting the mug shots it has integrated into its genome to neighboring cells. Because only the genomes of human sperm and egg cells get transmitted to the next generation, you could also not transfer the information about past aggressors to your children. Furthermore, while a mug-shot database can tell you about potential re-offenders, it does not help the unfortunate individuals first faced with an infection. It takes vastly more effort to assemble a human body than to construct a bacterial cell, so your body cannot afford to die at the first sight of a new danger. You need a system that launches counterattacks quickly against new threats and immediately spreads these counterattacks through your body. Your own immune system, and that of all animals with spines—the vertebrates—divides tasks among a collection of specialized cells. As in the case of bacteria, the crucial issue is identifying intruders. Employing strategies akin to those used by casinos and bacteria, your immune system creates molecules specific to each threat. It would not be possible to store one matching sequence for each possible intruder—the number of genes needed to do that would be larger than the number of letters in your genome. Instead, your immune system possesses a random mug-shot generator.

We saw that bacteria hunt down intruders by exploiting the way complementary DNA strands attract each other. Your immune system uses a similar strategy, though its mug shots, the antibodies, are based not on DNA, but on protein sequences. To understand how random antibodies are generated, let's take a closer look at proteins and how they are produced.

Your proteins are long words written in an alphabet comprised of twenty structurally similar molecules called amino acids. To produce a protein, these amino-acid molecules must first be assembled into a long chain. Then the newly made protein folds into a three-dimensional structure whose shape is based on the amino acids' physical and chemical properties (such as size, electric charge, repulsion of water). Each shape evolved, by natural selection, to perform a specific function (Figure 2.6). Because proteins are made up of an alphabet of twenty chemically similar molecules (letters) instead of four,

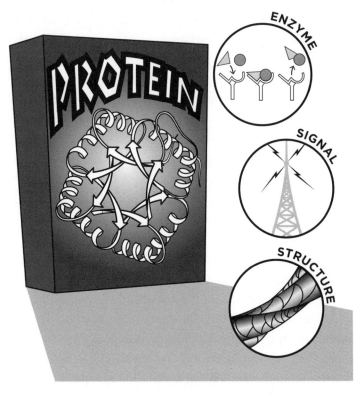

Figure 2.6: Proteins have diverse functions. A protein might catalyze a chemical reaction in which two specific molecules are encouraged to bind to each other by the way they fit into the protein's grooves (top right). Another protein might transmit information by passing around a high-energy chemical group (middle right). A third protein might join an assembly of like proteins to make a tiny support beam that helps uphold the structure of the cell (bottom right).

as in the case of DNA or RNA, they have a much wider range of potential structures.

For a cell to make a protein, it needs to translate a DNA sequence (with a four-letter alphabet) into a protein sequence (with a twenty-letter alphabet). With only minor variations, this translation follows the same rules across all the many life forms on earth; this is one of the reasons we believe that life arose only once on this planet. If you were to design such a translation scheme yourself, you would find that you need DNA words of at least three letters per amino acid: with two-letter words in a four-letter alphabet, the most you could specify would be $4 \times 4 = 16$ of the 20 different amino acids. Cells indeed use three-letter words (called codons) for this purpose: AAA, AAC, AAG, AAT, . . . , TTT. Altogether there are $4 \times 4 \times 4 = 64$ different codons to specify twenty amino acids. Thus, the code is redundant: most amino acids are encoded by more than one codon. For example, TGT and TGC both encode the amino acid cysteine. This redundancy is not random. The way the codon translation table evolved minimizes the impact of "typos" introduced during transcription.

To make proteins, the cell follows a succession of steps known as the "central dogma" of biology: the flow of information from DNA to RNA and then on to protein. The DNA sequence of a protein-coding gene has an average length of 1,000 letters. The sequence is copied into an RNA messenger by the polymerase we met in Chapter 1. This messenger RNA is then fed into the ribosome, another protein assembly. As it traverses the RNA sequence, the ribosome adds the matching amino acid for each three-letter codon to the growing protein (Figure 2.7). Cells inherit some copies of the proteins that form polymerases and ribosomes from their mother cells, so that the process of protein production could begin.

The antibody proteins that help recognize pathogens are Y-shaped. Each one is able to bind a certain specific group of fragments from intruder proteins to the two tips of its Y. But if

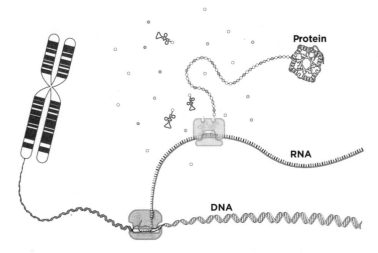

Figure 2.7: The "central dogma" of biology: a polymerase copies (transcribes) DNA into messenger RNA, which is translated into protein by a ribosome. The trumpetlike figures, called transfer RNAs, supply the matching amino acid to the ribosome.

each protein—and thus each antibody—is unambiguously specified by a corresponding gene sequence, how could the immune system make random antibody proteins? You may be familiar with a game called "Mixies," which involves assembling a picture of a human body from cards that depict a head, a torso, or legs. If the cards consist of twenty versions of each of the three body parts, you can generate thousands of different composite bodies. This is what your immune system does to generate antibodies: instead of storing preformed mug shots, your immune cells assemble a wide variety of mug shots from a limited set of components (Figure 2.8).

In most of your body's cells, the antibody gene is not specified in the genome, at least not in finished form. Instead, the antibody gene is assembled anew as your body produces

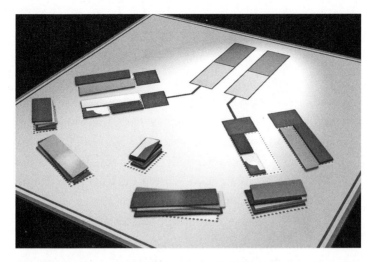

Figure 2.8: In the VDJ system, variants of different gene sections (the stacks of cards) are assembled into a wide variety of antibodies.

B-cells, the immune-system cells that patrol the antibodies around the body, using them to detect intruders. On each of three of your chromosomes you have neighboring regions called *v*ariable, *d*iverse, and *j*oining: collectively, the VDJ system. Analogous to the sets of head, torso, and legs cards in the Mixies game, each region contains a set of different versions for one section of the antibodies. When your body produces a B-cell, protein assemblies inside this cell edit the genome, using just one "card" from each section. The three cards are pasted together, forming a newly shuffled antibody gene. It is as if the genomes of your other cells contain the whole Mixies card set, while your B-cells assembled one particular figure and threw out all the other cards. This is unusual behavior: your B-cells are among the very few cells in your body allowed to edit their own genomes.

Before being approved for release, the B-cells' antibodies are compared against anything your body produces. Just as a casino guard entrusted with a computer-generated catalog of all possible faces would first have to throw out those that resemble law-abiding customers, your immune system first has to filter out any antibodies that could bind against your own proteins. If such self-binding went unchecked, B-cells would regularly initiate attacks on your own body in the same kind of autoimmune reaction that occasionally occurs in the bacterial CRISPR immune system. During its maturation in your bone marrow, any B-cell encoding an antibody that binds your own proteins commits suicide. The remaining B-cells are released into the body, hunting for intruders. When the B-cells find enemies, they then call on eater cells to execute them.

What Would Darwin Do?

The B-cells go about their hunting business by patrolling your body and reading off status reports from the surfaces of other cells. The status report originates within each cell, where proteins that have come to the end of their life cycles are broken up into pieces that float around. A specialized troop of proteins picks up these pieces and brings them to the cell surface, exhibiting them to the outside. These displayed fragments now provide a sampling of the kinds of proteins that are found in the cell. Mostly the fragments are from proteins produced from your own genes, and the status report reads "All is well." However, when an intruder is active inside the cell, some fragments of its proteins will be among those presented on the cell's surface—the cell's equivalent of screaming

"Help, I've been invaded!" By binding itself to that cell, the B-cell signals that a potentially dangerous foreign agent has been caught.

But that's not the end of it. The number of B-cells in your immune system is limited, so even though your VDJ system is, in principle, capable of making about a million million antibodies, you cannot make a B-cell for every possible foreign protein fragment. Instead, every B-cell binds to a range of slightly different fragments. The binding may be weak for some of the fragments, but your immune system has a way of strengthening that binding, thereby allowing it to launch a full and long-lasting defense. This is achieved through two brilliant measures that invoke the power of natural selection.

First, the B-cell with the matching antibody is given a prize: a signal to multiply itself, to spawn more B-cells that contain the successful antibody gene sequence. With more clones of this B-cell patrolling the body, cells infected by the matching intruder are more likely to be identified and eliminated.

The second, connected, measure ensures that the B-cell binds the intruder's protein fragments strongly enough to weed them out. How is this binding optimized? We have already seen that a population (be it a species or a collection of cells) can adapt by natural selection if there is heritable variation that affects the ability to spawn offspring. To exploit the logic of natural selection, the immune system has to create heritable variation among B-cells in the ability to bind the intruder's proteins and make sure that the best-binding B-cells multiply faster than weaker binders (Figure 2.9). If this is achieved, the immune system inevitably enriches itself with B-cells that bind better to the current intruder.

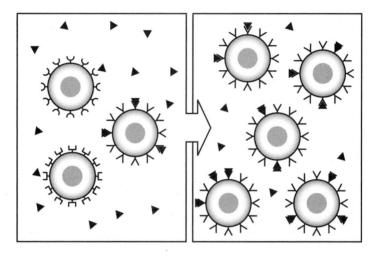

Figure 2.9: Evolution in B-cells. B-cells hunt for intruders using antibodies on their surface. The B-cells that most strongly bind parts of the intruder are selected to replicate.

When the signal to multiply is issued to a moderately successful B-cell, a unique program kicks into action. This program intentionally introduces mutations into the part of the gene that encodes the tips of the Y-shaped antibody protein. These hypermutations are programmed to introduce just the right amount of variation: about one new mutation in every 1,000 cell divisions. Hypermutation creates a huge diversity of B-cells, binding the intruder's protein fragments with variable strength. Note that it is not the B-cell's whole genome that is hit by mutations, but only the part of the genome that determines the binding specificity. This process is also highly unusual: no other region in your genome is ever deliberately mutated.

By chance, some of the mutated variants will bind the intruder's protein fragments more strongly than the original

B-cell antibodies did. The cells that manage the immune system will spur these improved B-cells to further multiply. After a few rounds of the cycle of hypermutation followed by the multiplication of improved B-cells, the immune system will be swarming with B-cells that bind perfectly to the intruder's proteins. Long-lasting survival is ensured for some representatives of those who win the B-cell competition. They are turned into memory cells, which stay in your body to protect you from future attacks by the same aggressor: now you'll have antibodies against that particular disease.

Darwin's three requirements of natural selection are perfectly captured by this system. B-cells vary in the antibodies they produce, both because of the VDJ system's Mixies-like shuffling and because of targeted hypermutations. Variation exists? *Check.* When B-cells are allowed to multiply, they pass on their particular antibody genes. Variation is heritable? *Check.* The signal to multiply is based on how well the antibodies bind to the intruders. Variation affects fitness, the ability to leave offspring? *Check.* With all three requirements of natural selection in place, your B-cells adapt to your pathogens as a logical necessity.

Technically speaking, because of the unusual way in which they are assembled, antibody genes do not belong to the society of genes. They are fleeting assemblies, varying from B-cell to B-cell. The society members—the alleles—of this system are the complete Mixies sets of the variable, diverse, and joining regions. These regions are entirely useless by themselves; they can defend the body only when they are employed by the machinery that cuts and pastes them into an antibody-encoding gene. This machinery is itself encoded in several other genes, and many more genes contribute to a functioning immune system. The arrangement of shared labor among these genes allows the whole society of genes to thrive in the presence of pathogens.

Double Agents and Baby Giraffes

Creationists often argue that something as complex as a human being could never have arisen by chance alone. Mutations are indeed random, that is, not biased toward changes that increase fitness. But the process of natural selection is the very opposite of random. As we saw in Chapter 1, cancer has a random component, as its progression relies on the generation of mutations. But cancers evolve according to the logic of natural selection, resulting in a faster rate of proliferation for the first cancer-promoting mutations. Similarly, the adaptation of your antibody repertoire is based on the generation of random variants. But, crucially, the proliferation of B-cells after infection is not random.

Darwin's evolutionary theory is remarkable in its simple logic and its ability to explain a wide array of biological observations. Darwin was not the first to ponder the mechanics of evolution. Born sixty-five years before Darwin, the French naturalist Jean-Baptiste de Monet, Chevalier de Lamarck, also championed the notion that species evolve over time and adapt to new environments. But Lamarck had a very different idea about what process underlies evolution.

Lamarck's idea, embraced by many of his contemporaries, was that a change acquired during a creature's lifetime is passed

Natural Selection (Darwin)

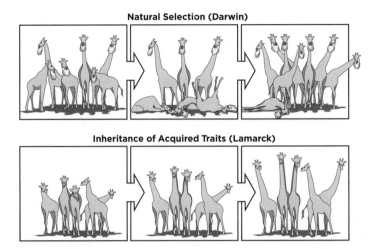

Inheritance of Acquired Traits (Lamarck)

Figure 2.10: According to Darwin, evolution proceeds by random variation and natural selection. If there is heritable variation in neck length, then taller giraffes find more food and thus are better at surviving—leaving more offspring that inherit long necks. In Lamarck's theory, the giraffes' necks get longer over their lifetime from continuous stretching to reach the higher leaves, and their elongated necks are inherited by their children.

on to that creature's children. This idea is nicely illustrated with a classic story of how giraffes might have evolved their long necks. Imagine a giraffe passing its days in the savanna. With all the low-hanging leaves of the area exhausted, the giraffe might stretch its neck to reach the higher leaves. From continuous stretching over many years, its neck gains a few centimeters in length. According to Lamarck's theory, the elongated neck would be inherited by the giraffe's children (Figure 2.10).

If Lamarck's ideas were correct, then any ability you acquire could be passed on to your children. If you learned to play the piano, your children might be born with that same level of skill—no necessity for expensive lessons or hard practice. This, of course, contradicts common experience, and consequently, Lamarck's ideas were quickly abandoned as the explanatory power of Darwin's theory became apparent.

We now know a great deal about how information is transmitted from one generation to the next through the genome. Each discrete bit of heritable information—each copy of an allele—is confined to one particular cell. In multicellular organisms, such as giraffes or humans, any change specific to the neck (or any other body part) can affect only the information stored in the cells of that part. That information cannot be transmitted to the genomes of ovaries and testicles, so it cannot be inherited.

According to Lamarck, changes do not occur at random. Rather, they come about through interactions with the environment, as in the story of the giraffe. In contrast, the modern interpretation of Darwin's theory assumes that there is a barrier between an organism's genome and its environment, that the passage from genes to the environment is a one-way street. The genes' products are tested by the environment for their effect on fitness, but the teachings of the environment never directly make their way back to the genome. If the environment demands a longer neck, this message is not transmitted directly to the giraffe's genes; those genes cannot "answer" the call by changing to produce longer necks. Instead, the giraffes in a population differ from one another in a number of mutations. Those whose necks are longer have easier access to food, and if the longer neck is indeed due to a specific set of mutations, then their offspring will thrive in time. The same Darwinian logic underlies the development of cancer and the action of your immune system.

Let's go back to the bacterial immune system and see how it fits into Darwinian logic. As explained earlier in this chapter, in the bacterial immune system, genomic variation is not randomly generated; instead it is drawn straight from the environment. When viruses in the environment invade the organism, they leave their traces in the bacterium's genome, and that acquired information is passed on to future generations of bacteria.

Once the viral sequences have been integrated into the bacterial genome, they switch sides, thriving within the bacterial society of genes and helping to fend off their viral cousins. The workings of the bacterial immune system are plainly Lamarckian, functioning not through random variations that are tried out in the environment but through a direct route from the environment (the virus) into the genome.

Does this mean that the logic of natural selection has somehow been superseded by a different mode of evolution in bacteria? While there is a clear Lamarckian component in the bacterial immune system, natural selection is as central to it as it is to all adaptations in the living world. Lamarck's ideas pertain to how the bacterial immune system functions, but they cannot explain how this system first arose in the evolutionary history of bacteria.

Even though we do not have a clear picture of the evolutionary steps that produced the bacterial immune system, there is no reason to assume that it arose in any way other than through heritable variation and natural selection. We can imagine that way back in time a population of bacterial cells had immune systems that were slightly different from one another due to random mutations in the genes that organize the

CRISPR system. Those bacteria with a more effective immune system would be better at surviving viral infections. Over time, they out-competed those with less efficient systems.

The bacterial immune system establishes a way to inscribe important memories into the genome, and this unique passage from the environment to the genome evolved by natural selection. Natural selection as understood by Darwin is powerful enough to underlie each and every adaptation that we see in the living world.

Lamarck's Milk

Our immune system, just like that of bacteria, retains memories of past infections. The genomes of our B-cells reflect the battles fought during our lifetime. The more battles we fight successfully, the richer our repertoire of memory cells becomes. When a child first encounters the measles virus, the immune system has to go through the whole VDJ / hypermutation / selective multiplication cycle to learn how to deal with it. The child's body keeps the corresponding memory cells, which confer immunity from the measles virus. That is why we don't get the measles a second time.

It is unfortunate that our immune system's genomic memories cannot be transmitted to our children. However, there is a nongenomic and quasi-Lamarckian path by which children can benefit from their parent's experiences with infections. Mother's milk not only provides nutrition tailored to the needs of human babies, it also contains many immune-related molecules, such as specific sugars that can prevent nasty bacteria from attaching to the infant's gut wall. When a mother has recently been exposed to a virus or bacterium, the antibodies she produced in response make up a large fraction of the proteins contained in her milk. These antibodies come in a form that is particularly resistant to digestion, so they stick to the corresponding bacteria or viruses in the infant's gut. Also, they hang around in the baby's mouth and nose, suppressing airborne diseases.

Mother's milk thus boosts the infant's immune system, reducing the incidence of colds, flus, and other diseases. This is one of the reasons the World Health Organization recommends exclusive breastfeeding for the first six months of life, with continued supplementary breastfeeding up to age two and beyond. Breastfeeding, the defining property of mammals, is a great strategy in the everlasting fight against pathogens.

As we have seen, one organism attacking another—whether it is a virus invading a bacterium or a bacterium attacking us—is essentially a clash of societies, fought by highly efficient, dedicated task forces. The human immune system harnesses the power of natural selection to fight enemies in real time. In evolutionary time, however, our bodies are only fleeting assemblies. The society of genes evolves from human generation to human generation, and the transition between the generations—with the checks and balances that regulate it—is the topic of the next chapter.

3

What's the Point of Having Sex?

In conflict, be fair and generous.
In governing, don't try to control.
—Lao-tzu

IN 2013, THE BANK OF ENGLAND REPLACED THE £10 NOTE'S portrait of Charles Darwin with a portrait of Jane Austen. Both are illustrious British figures, but beyond that, you may think they do not have much in common. However, if you look more closely, you will see that their work has a common subject: they both wrote about sex. The task of Austen's female protagonists is finding a suitable mate. Put in Darwinian terms, these women are searching for partners whose genes they will combine with their own, mates who will be advantageous to their offspring both genetically and socially. Sexual reproduction is a crucial driver of the society of genes' evolution. Darwin knew nothing about genes, but he understood the importance of sexual reproduction. If he were alive, we'd like to think he would have been happy to cede his spot on the £10 pound note to Austen, his colleague in spirit.

The Advantages of Sex: Beyond the Obvious

In the cases of cancer and the immune system, the society of genes evolves in a single person. The body's cells form a massive population that adapts according to the principle of natural selection. But despite all that adaptation, this population of cells will inevitably die out after at most a few dozen years. The only way our genes can survive is by making copies of themselves that pass into the next generation, that is, to make babies. To our genes, it does not matter whether babies are conceived in the conventional mammalian way or whether the ever-expanding possibilities of reproductive medicine are involved. We will adopt that stance and use the word *sex* when the genes of two individuals are mixed to create a new genome.

Is sex really a good strategy to make children? To explore this question, let's first look at what fathers and mothers bring to the table when they conceive a child. To have a child, your father had to accept that he could copy only half of his genome into yours. This contribution is halved further in each generation: his grandchildren will each have just a quarter of his genome, and his great-grandchildren will have only one-eighth. Within just fifteen generations, your father's 20,000 genes will be diluted down so much that his descendants will on average inherit not more than a single one of his alleles.

What if mothers could instead go ahead with the reproduction business all on their own? This may seem far-fetched, but it actually does happen in some animals. Rather than seeking the DNA of a "collaborator," the mother would put her complete genome into the egg and produce clones of herself. Her cloned daughters would produce cloned grandchildren. In fifteen generations, all your mother's descendants would still carry her complete set of alleles. No sex, no genetic dilution.

As this argument shows, to put the same number of alleles into the next generation, sexually reproducing individuals have to produce twice as many children as those who reproduce by cloning. Sex, it seems, comes at a high price; a burden known as the twofold cost of sex. To pay such a tremendous price, there must be an equally tremendous compensatory benefit. And since in many cases all that the male contributes is half his genome, the genome is where we should look for answers.

To paraphrase an old joke, a male fashion model once met a female physicist at a cocktail party. "Let's get married," he says. "Our kids will be as pretty as me and as clever as you." She replies, "But what if it's the other way round?" If the couple's intelligence and looks were each determined by a single allele on one of their chromosomes, both outcomes would be equally likely. Sex does not involve a merger of two complete genomes; if it did, genome size would double every generation, a logistical impossibility. Instead, a child's genome will always have obtained only half of mom's and half of dad's. The child's brains and looks thus depend on exactly which half was chosen from which parent. The power that sex holds for the society of genes is rooted in this random mixing of alleles.

Not all organisms use sex to replicate. Bacteria do not have sex, at least not sex as we know it (more on that later). To spawn offspring, bacteria duplicate their genomes and produce clones of themselves. The genomes of mother and daughter bacterium are exactly identical, except for a few incidental mutations. No costly dilution here.

The following experiment demonstrates how bacteria evolve. You create a miniature football field inside a dish (Figure 3.1 shows one half of it) and cover the bottom with a sugar solution that bacteria love to eat. In the regions beyond the goal lines, you lace the solution with antibiotics at increasing concentrations: 1×between the goal lines and the 10 yard lines, 10×between the 10 and 20 yard lines, 100×between the 20 and 30 yard lines, 1,000×between the 30 and 40 yard lines, and 10,000×between the 40 and 50 yard lines. You then sprinkle bacteria evenly onto the field and wait to see what happens. First, you will see that bacteria are growing only at the antibiotic-free end zones. Soon there will be trillions (that is, millions of millions) of cells behind the goal lines.

Eventually, some patches of bacteria will start to grow beyond the goal lines, into the region contaminated with antibiotics. As in the evolution of cancer, each of the trillions of cell divisions has a certain probability of introducing one new mutation into the corresponding genomes. After trillions of cell divisions, there will be so many different genomes present that, by chance, a few will have mutations enabling them to venture into the territory contaminated with antibiotics. As in cancer, there is power in large numbers.

Out of the patches of lucky mutants, bacteria will spread until they have conquered the whole goal line to 10 yard line regions. However, because there is a tenfold concentration of antibiotics beyond the 10 yard lines, they are not able to continue their growth into those regions. For this encroachment, an even more refined skill set is needed, requiring additional mutations. But again, the new genomes that spread up to the 10 yard line achieve power in numbers, with one eventually

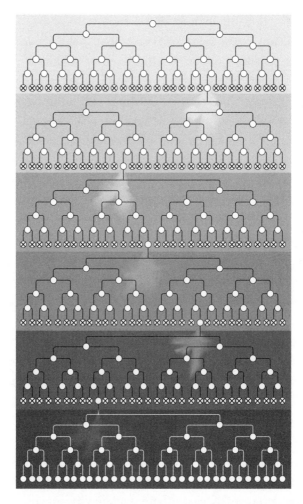

Figure 3.1: An experiment performed by Roy Kishony and his colleagues at Harvard Medical School. A large rectangular dish filled with a sugar solution is divided into sections resembling the yard lines of a football field. The figure shows half of the field, starting at the end zone at the top and ending at the 50 yard line at the bottom. There is no antibiotic in the end zone region at the top, and a low concentration of antibiotic in the 0–10 yard region. Antibiotic concentrations increase tenfold beyond the 10, 20, 30, and 40 yard lines (top to bottom). When bacteria are thinly spread over the whole field, they grow in waves: first only up to the goal line, then up to the 10 yard line, and so on. The transitions across each line require additional mutations, which become likely once there are enough bacteria with the previous mutations.

hitting on the mutation necessary to conquer the next sector of the field. This process of expansion, mutation, and power in numbers continues until the bacterial populations that started from both ends of the field meet up, finally competing with each other for the sugar left in the middle of the field. As happens in cancer, the bacteria acquired several mutations, each allowing them to cross one line of defense, and this progress was accelerated massively by the proliferation that ensued after each mutation. The stepwise path to increasing antibiotic resistance is as scary as the stepwise path to cancer: infections by multidrug-resistant bacteria are growing much faster than the rate at which pharmaceutical companies can develop new antibiotics.

The reproduction of bacteria, in which each daughter cell is genetically identical to its mother cell (save for, at most, a few mutations), has the following crucial limitation. Several bacteria are likely to independently acquire different mutations, each of which may increase antibiotic resistance. However, the descendants of only one of these mutations will eventually outcompete those of the other cells, thereby firmly establishing that mutation in the bacteria's society of genes. The competing mutations will be pushed out of the competition (Figure 3.2).

If bacteria engaged in sex, the acquisition of antibiotic resistance would have been stepped up even more. In that case, each favorable mutation could benefit from joining forces with the others by being combined into the same genome. The daughter cells of a union between two bacteria with complementary mutations would be even fitter than any of the parents by themselves and might immediately be able to venture beyond the next line of the football field.

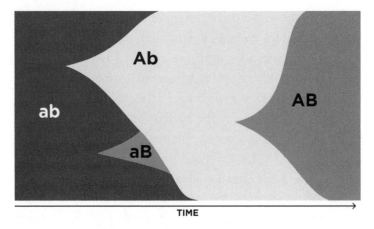

TIME

Figure 3.2: A hypothetical evolutionary scenario in a bacterial society of genes, focusing on two genes that each can come in a low-fitness allele (a and b) or in a high-fitness allele (A and B). The arrow indicates the direction of time. Each vertical slice of time shows the distribution of alleles at a given moment. Initially, all bacteria have the low-fitness alleles a and b. The high-fitness alleles A and B arise independently and coexist in Ab and aB individuals for a while. But because there is no sex and no recombination, the two mutations cannot be merged into one genome, and aB is pushed aside by Ab. After a long time has passed, the B mutation reoccurs in an Ab individual, finally creating bacteria that combine the two favorable mutations.

Without sex, the competing advantageous mutations will be lost. Similar mutations may or may not occur again in a future generation. By shunning the twofold cost of sex, the bacterial society of genes instead pays the price of not being able to combine good mutations that arose in different genomes. Bacteria still do a good job of adapting to changing environments, drawing power from their large numbers. But in a population as small as our own species, too little sex would have destined us for extinction as soon as fast adaptation to a changed environment became a matter of life and death.

All mammals pay the price of genome dilutions that comes with sex, but, in turn, they are able to combine advantageous properties from two parents into one child. As unromantic as it sounds, sex not only provides a way for good alleles to come together, but it also provides an efficient way to purge harmful mutations from the society of genes. Imagine a couple in which both parents' genomes contain harmful mutations, each affecting different genes. If each parent reproduced as clones, all alleles in their genomes would be doomed: over time, their clones, burdened with bad mutations, would lose out in the struggle for life. If, however, there was a way to combine all the intact alleles of both parents into one child, leaving out the harmful mutations, the remainder alleles would have saved their necks—they would have managed to disassociate themselves from their burdensome genomic neighbors.

This is the point of sex. Sex makes it possible for alleles to live the American dream. It unlinks them from one another, so that even if a favorable mutation appears in a bad genomic neighborhood, that allele still has a chance of succeeding. As we shall soon see, the allele can find a new neighborhood and slowly rise to popularity, while the bad alleles from its original neighborhood will not do as well. Essentially, sex evolved in the society of genes because it allows the society members to continuously form new alliances and hence to work together more efficiently in the long run.

Here's another way to think of it. Card games such as bridge are played in teams. If these teams are fixed, then the success of a player is critically dependent on the skills of her partner; a brilliant player teamed up with a bad partner is unlikely to do well. However, if teams are randomly reassembled in each round, it is the quality of the individual player that determines her total score. Similarly, the society of genes assembles a wide variety of allele teams into genomes through sex, so that natural selection will, in the long run, promote the best-performing alleles.

Sex Is Egalitarian

For each matching pair of alleles on the two chromosome sets you inherited from your own parents, only one can go to your child, while the other has to stay behind. To ensure fairness, species that use sex have a special method of cell division. Called meiosis, this process is at the very heart of sexual reproduction. If chromosomes were transmitted from generation to generation in one piece, the mixing of genes would be severely limited. For example, the fate of all of the alleles for about 4,000 genes inhabiting chromosome 1 would be tied together for eternity. Let's say the physicist in the joke about the couple at the party drew her ability for formal reasoning from an unusual allele she inherited on her mother's chromosome 1, and her ability for creative thinking from an allele of a different gene on her father's chromosome 1. If she could transmit only one of these chromosome 1 copies, the two abilities could never make it into her children together.

How is a fair mix of each parent's chromosome 1 produced? The cellular machinery starts by making a second copy of each of the individual chromosomes. The matching chromosomes 1 are then lined up, and pairs of them are cut into two or more corresponding pieces. New chromosomes are assembled by choosing one of these pieces for each region. The choice is not guided; a random molecular dance determines how the DNA

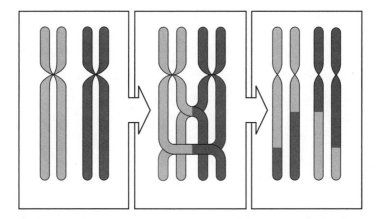

Figure 3.3: After their duplication in preparation for recombination, the maternal and paternal chromosome copies are present in two linked copies each. They then recombine, swapping matching regions.

will reassemble. This crucial genomic preparation for making egg and sperm cells is called recombination (Figure 3.3). To borrow the analogy Richard Dawkins introduced in *The Selfish Gene*, picture the maternal grandfather as having a full deck of blue cards and the grandmother having a full deck of red cards. When these decks are recombined in their daughter's egg-cell production, new decks are generated, each with the full fifty-two cards, but with the blues and reds shuffled. The point of sex is to shuffle alleles into new genomic combinations, and recombination achieves this goal.

All chromosomes undergo recombination, with the exception of the Y chromosome. The Y chromosome determines the sex of the embryo, through a gene called SRY: boys result when the gene is present and functional, girls when it is absent. An XY chromosome set can result in a female embryo if the SRY gene is broken or is not recognized by the proteins with which it needs to interact. Because males have one copy of the Y and females have none, the Y never meets a matching chromosome—that is, it never has a chance to shuffle its alleles with the alleles on other Y chromosomes. The exception is a region encompassing about twenty genes in which the Y chromosome mirrors a matching region on the X chromosome. When the chromosomes pair up in male meiosis, X and Y join up in these matching regions, recombining in the region just as do the other twenty-two chromosome pairs. But the alleles on the rest of the Y are condemned to eternal companionship. Following the deck-of cards-analogy, this is as if only a small stash of cards on the top of the Y deck would ever get exchanged with cards from the X deck, leaving the rest of the stack unshuffled.

If a Y chromosome suffers just one seriously damaging mutation in its nonrecombining part, all the alleles on that chromosome are doomed: there is no way to get rid of the mutation. If, however, the mutation is only moderately damaging to its carriers, it might hang in there. Without recombination, there is no way to purge individual harmful mutations from Y chromosomes, and so Y chromosomes are slowly decaying.

We have good evidence that about 150 million years ago, the X chromosome and Y chromosome were a matching pair, just as our other twenty-two pairs of chromosomes are now. Slowly, an asymmetry developed between these two chromosomes, and they started to diverge. Today, while the X chromosome is still home to about 2,000 genes, the Y chromosome contains fewer than 200. Over the many million years without a partner to pair up with, mutations have irreversibly wiped out one gene after another on the Y chromosome. Meanwhile, equally destructive mutations that occurred on the X chromosome were efficiently weeded out by recombination with the second X chromosome in women. Ironically, the chromosome that determines an embryo's sex is the only one that suffers from getting no sex at all. If humanity holds out long enough, in millions of years the Y chromosome might disappear altogether. Males might then be specified by the lack of a second X chromosome. This has already happened for the spiny rats that live on some Japanese islands; they do fine without Y chromosomes.

Getting back to the shuffling: once a woman's chromosomes have recombined into fresh combinations, one of the new chromosomes from each pair has to be inserted into the newly made egg cell. From the point of view of an allele, getting into the egg is the only chance it has at making it to the next generation. Any allele would become extinct if it consistently failed to appear in egg and sperm cells. Given these high stakes, it is amazing that meiosis is fair—but it is. If there were any loopholes in how places in the sperm and egg cells are assigned, the society of genes would teem with the most successful cheaters instead of the alleles most useful for the survival of the society as a whole.

In meiosis, every allele gets a 50:50 chance of entering a particular shuttle to the next generation. It's not deterministic: meiosis is like a coin toss. Chance plays an important role in the fate of alleles, as we shall see again in Chapter 4. Crucially, meiosis is blind to the quality of individual alleles. The assembly of variation through recombination is random—nonrandomness comes later, in the form of natural selection.

Instead of a coin toss, wouldn't it be better if a mother's cells decided which allele was allowed into the egg based on merit, always selecting the better out of two matching ones? Let's assume for a moment that for a specific gene, the mother has one nicely functioning allele on the chromosome inherited from her mother and one that is defective on the one inherited from her father. Instead of always choosing the functional allele, meiosis provides an equal chance for the defective version to be passed on. This does not sound like an efficient system.

But in the alternative, who would decide which copy is better? As we shall see in Chapter 5, the consequences—and hence the "quality"—of an allele depend largely on the versions of other genes it cooperates with in a given genome. Thus, even if an alternative meiosis machinery would be able to tell which of two competing alleles works better in the current genome, it could not accurately predict which allele would fare better in the next generation. In the same vein, why should every citizen have the right to vote? What if only some citizens were allowed to vote, based on their moral superiority? The problem, of course, is defining moral superiority. And how could there be an infallible judge of these qualities?

Possibly worse than the uncertainty associated with meritocracy is that this type of system provides its members with ways to cheat: if someone or something decides who is deserving, that decision might be influenced. History suggests that a functioning, egalitarian democracy trumps all other forms of government in terms of consistently furthering a society's overall welfare. So, instead of trying to award the best genes with a place in the shuttle to the next generation, the society of genes gives all its members equal rights.

The number of possibilities enabled by sex is astounding. To get a feel for just how amazing it is, picture a population of organisms with genomes that have only 1,000 genes, each one of which occurs in two different alleles, which we call "A" and "B." A particular genome half might contain allele A of the first and second genes, allele B of the third gene, and so on. How many different genome halves could there be? It's an easy calculation. The first step is to sort the possible genomes into two groups, according to which allele of the first gene they have. In each of those two groups, there are two groups depending on the second gene. In total, there are 2×2 groups distinguished by the first two genes. Extending this logic, you end up with 2×2×2 . . . ×2—that is, 2 multiplied 1,000 times with itself, which is a quantity much larger than the number of atoms in the known universe. And remember that 1,000 is a small number of genes—your own genome has 20,000—and that two alleles are many fewer than the variability seen in real genomes. Each of your genes exists in tens, if not hundreds, of alleles across all of humanity, as we shall see in Chapter 4.

Even without considering new mutations, the mixing of already existing genomes can produce an astounding array of new variations. Through recombination, human populations sample combinations of variants in numbers far beyond the range available to species that do not have sex. New genomes with variants that do not work well together will not be successful. Other combinations will work out extremely well, such as that hypothetical beautiful and smart child of the model and the physicist.

The organization of the society of genes closely resembles the guild system that organized crafts and trades in medieval

European cities. Each guild had strict rules about what their members were allowed to produce and which tools they could use. These requirements enforced clear delineations among guilds. We can say that each genome corresponds to an assembly of craftspersons with exactly one member from each guild. Thus, recombination and sex do not pile 20,000 random alleles from the society of genes into one genome. A specific place on the chromosome is always inhabited by alleles for the same gene—craftspersons from the same guild. The crossing-over of chromosome arms in recombination is carefully orchestrated to ensure that parts of a chromosome you inherited from your mother are replaced with exactly matching parts from your father's chromosome. Thus, the organization of the chromosome into "guilds" keeps the order and location of the genes intact.

Despite knowing all the advantages of sex, perhaps you find the idea of producing clones of yourself appealing. But consider the consequences of a mutation that causes its carriers to reproduce clonally. In the short term, this newly created allele would do very well indeed. Your genome specified a person with the intelligence and good taste to read this book, and so your clones and grandclones would inherit a worthy set of genes. The problem would crop up in the long term. The only variation that would occur in your descendants is that generated by occasional mutations, errors made during the cloning process. But, without noticeable variation, your future clones would be at a severe disadvantage when facing crises. They would be slow to adapt to a sudden change in the climate, for example. The necessary mutations would have to occur one after the other in one clonal lineage of mothers and daughters; there would be no way to combine beneficial mutations that arose in separate individuals. Eventually, your clone population—now separated from the remaining society of human genes—would likely arrive at an evolutionary dead end and face extinction.

This is no hypothetical scenario. Quite a few animal species do without sex, among them certain species of sharks, snakes, and insects. None of them seem to hang around for very long. They die out fast, with the cloning strategy rarely surviving for more than a few million years. Almost all clonal animal species we see are relatively recent experiments of nature, trying out what happens when a species refuses to pay the twofold cost of sex. When their environment changes faster than they can adapt to it by means of individual mutations, they are in trouble. No mammals have ever been observed to produce offspring without engaging in sex. One reason for the success of our particular group of animals may thus be that we fortified our reproductive system against the temptation of a cheap but ultimately fatal sexless life.

Why then haven't bacteria, which are sexless, become extinct? They draw power from numbers, as we saw in the football field experiment. Moreover, while bacteria do not have sex as we know it, they have found other ways to exchange genes with one another. As we shall see in Chapters 6 and 7, bacteria can mix and match their genes; they just don't do it in the highly ordered fashion used by sexual species.

There is one animal—known as the bdelloid rotifers—long believed to have lived an entirely sexless, female-only lifestyle for many millions of years. Now we know that these females are not entirely sexless from a society of genes perspective. They use a genome-mixing strategy akin to that of bacteria. So, genomically speaking, isolation indeed appears to be a slow-acting poison.

Big Stakes, Big Cheating

Meiosis is a fair process, but, as is typical in the life sciences, there are interesting exceptions. If an allele consistently gets into more than 50 percent of the sperm or egg cells of its carriers, it will fare better in the society of genes. As it turns out, there are indeed some successful cheaters in the game.

One example for such cheating is responsible for a disease called "achondroplasia," affecting one out of 20,000 newborns. People born with achondroplasia have a problem converting cartilage into bone, resulting in a shortening of arms and legs; as adults they are typically about 130 centimeters (4 feet 3 inches) tall. The reason is almost always a mutation that occurred during the production of sperm in the father's testicles: an exchange of one specific letter in the *FGFR3* gene, which encodes a sensor for growth signals.

Achondroplasia should be far more rare than it is, given the typical rate at which mutations occur. On average, less than one new mutation arises across all of the genome's 6 billion letters per cell division. Even when considering the many rounds of cell division in sperm production, this mutation rate should result in less than one achondroplasia case in a million births. So what explains the much higher incidence of this condition? Is the single letter responsible for the disease par-

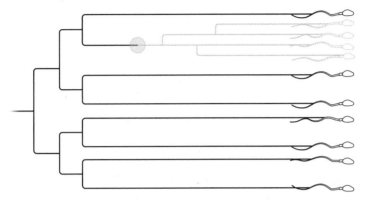

Figure 3.4: An example of natural selection in the production of sperm. A new mutation (the gray circle) that increases the rate at which cells multiply during the production of sperm will ensure that the corresponding allele is deposited in more sperm than its competitors.

ticularly unstable, causing it to mutate much more frequently than the other letters in our genome?

The explanation lies elsewhere. It turns out that this single-letter mutation not only affects cartilage but also increases the speed at which the cell lineage carrying the mutation divides when sperm cells are produced (Figure 3.4). Due to their faster proliferation, cells with this change contribute up to 1,000 times more to the mature sperm than other cells do. This is another example of natural selection acting inside the human body: for the cells in the testicles, just as for the creatures out there in the jungle, fitness is measured by how many offspring they produce. By outcompeting its equals, a new achondroplasia-causing mutation effectively rigs the normal odds, increasing its chance of making it into the next generation. This is a special case of natural selection: the newly arisen mutation can increase its odds of survival into the next generation, but not beyond. In a person born with achondro-plasia, all sperm cell precursors proliferate at the same fast rate, and there no longer is any advantage for the mutation.

Other ways to sabotage the egalitarian paradigm of sex are more insidious. We know that utterly selfish exploitations occur in the society of genes of fruit flies, and there is reason to believe that such genetic fraud also operates in the human genome. One of these exploitative systems involves two part-ners in crime, genes sitting next to each other on a chromo-some. During the production of sperm, one of the genes in-structs the cellular machinery to make a poison; the second provides the cure. The poison is exported outside the cell, and the cure remains inside. The poison is exported outside the cell, and the cure remains inside. The gene pair ensures its spread by using the poison to kill all sperm cells that do not also have the poison / antidote gene pair (Figure 3.5).

The devious gene pair contributes nothing to the survival or the fertility of the individual who possesses it. It doesn't help build blood vessels, improve the brain, or fight nasty bacteria. A male fly that inherits this pair of genes is in no way better off for having them. On the contrary, it pays a high price by killing off many of its own sperm. The pair damages the chances that the fly's genes will make it into the next genera-tion, but still the two genes profit. By getting into the majority of sperm cells, their own chance to make it into their carrier's children is far above 50:50.

Other kinds of trickery exist and have important repercus-sions. Remember that, in contrast to single-celled bacteria, which pass every mutation on to the next generation, only mu-tations in your parents' cell lineages that led to the produc-tion of sperm and egg cells—the germ line—had any chance

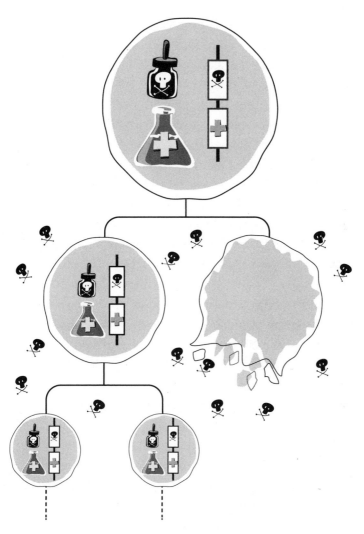

Figure 3.5: A poison / antidote pair of selfish genes. The poisonous protein and the antidote are produced in sperm cells that carry the gene pair in their genomes. Only the poison is exported to the outside of the cell, where it kills all neighboring sperm cells that lack the ability to make the antidote.

of being inherited by you. If you started life with anything fundamentally new—an allele that neither of your parents had—this novel allele must be due to mutations that arose in your mother's or your father's germ line.

You received roughly sixty such new mutations. Most of these were simple copying errors, accidental exchanges of one genomic letter for another. You might be surprised to learn that your parents differed in their generosity: your father endowed you with much more genetic novelty than your mother did. This is due to a fundamental difference in germ line cell production. Cells in the male germ line proliferate at an extraordinary rate: over his lifetime, a man produces hundreds of billions of sperm cells. In contrast, a woman produces a total of only a few hundred egg cells. By the time your father was twenty-one, each sperm cell had already undergone about 300 cell divisions in its production, and it didn't stop there. At the same age, your mother's egg cells had seen just twenty-two divisions. All her egg cells were essentially preproduced before she was born, so this number of cell divisions did not increase for egg cells released later in life. More cell divisions in men inevitably mean more copying errors, which explains why you can attribute the majority of your new mutations to your father.

Sperm production from a fixed set of cell lineages begins at puberty and continues until men die, year after year, and more and more mutations accumulate with each round of cell division. For this very reason, many genetic diseases increase in frequency with the age of the father. One example is Marfan syndrome, which affects about one in 5,000 people. This disease is caused by defects in the *fibrillin-1* gene. The gene is responsible for a fiberlike building block that plays an important role in assembling your connective tissues; these are crucial to constructing body parts as diverse as bones and heart valves. People with Marfan syndrome are unusually tall and have long, thin fingers. They often suffer from problems with their lungs, eyes, and major arteries. A substantial proportion of Marfan cases are caused by new mutations that arose in the father's testes. Because of this, children fathered by a man in his fifties are almost ten times more likely to suffer from Marfan syndrome due to a new mutation than those fathered by men in their early twenties.

Indeed, there is only a slight chance that a mutation will be good for the society. Humans, like practically all species, are reasonably well adapted to their environment. Thus, most new mutations—if they have any noticeable effect at all—tend to reduce fitness rather than increase it. If a man cared about reducing the mutational load he passes to the next generation, he should have children in early adulthood, when fewer mutations have accumulated in their sperm.

This Is Not about You

In understanding sex, we have taken the perspective of individual genes, not the perspective of the men and women whose genomes harbor them. As Richard Dawkins wrote in *The Selfish Gene*, the individual is a fleeting assembly of molecules, whereas genes and their alleles can last for millions of years and more. It is the genes that skip down the generations by maneuvering us, their "survival machines." That name is not meant to suggest that our genes want *us* to survive. Our genes engineer us to last not much longer than the time it takes to have produced enough children to carry them into the following generation.

The democracy of sex is the agency for the shuffling of the deck, the re-sorting of the alleles. The mechanisms of sex are set up by proteins encoded by many genes. These have earned their place in the society of genes because they provide a helpful service: sex unlinks the genes from one another, so that in each generation, many different combinations of alleles are tested against the environment. With sex, each allele can essentially go it alone, one generation at a time.

You might counter that it is you, the individual, who lives or dies before passing on the genes—thus, shouldn't natural selection occur at the level of the individual? Say you have an allele on one of your chromosomes that is also present on one of the chromosomes in 1,000 other people. Half of the children of these people will inherit the allele, and so the allele's fate hinges on their total number of children. If those with the allele on average have more children than those with competing alleles, the allele will continue to spread and prosper. From the point of view of the gene, however, it is irrelevant who of those specific individuals have children. Each person is a single battle in a larger war that determines the frequency at which the allele will spread. An allele that caused half of its carriers to die young and the other half to have four times the average number of children would generate twice as many offspring as its competitor. That allele would do very well indeed, even if it spelled doom for half of the people inheriting it.

The Genomic Battle of the Sexes

The differences in sex chromosomes are only the starting point in understanding differences between women and men. While all genes except those on the Y chromosome are inherited both by males and females, many of those genes are active only in males, while others are active only in females.

The primary reason for many of the differences between the sexes lies in the fact that females generally invest much more in their children than men do, at least initially. The mother makes an egg, a big repository of nutrients that supports the early embryo. The father's sperm is optimized for just one task: the efficient delivery of his DNA into the egg cell. The initial imbalance between parental investments continues with the nine months the embryo is cared for in the mother's womb, and traditionally culminates with the mother breastfeeding the baby for many more months. Because of this imbalance, men and women evolved different instinctive strategies. Ultimately, since the mother invests more in her offspring, she has reason to be more selective in her choice of a mate. After all, it is much more costly for her to correct a bad decision. Consequently, if an allele arose that produced choosiness in the women it inhabited, that allele would increase its carrier's fitness and thus likely establish itself in the society of genes.

The different strategies of men and women can lead to severe genomic conflicts between the mother and the father, with the embryo providing the battleground. When the mother produces smaller and weaker babies, she preserves some of her own resources, increasing her chance to survive childbirth and live on to produce more offspring. However, smaller and weaker babies have an increased risk of not surviving to adult-hood, so the mother's investment represents a compromise. What does this look like from the perspective of the father's genome? It is always possible that the mother would later have children with a different mate. Given that consideration, the father's genes would fare better if his child took more of the mother's resources now; this would increase the chance of the child's success (and the success of the father's genome) at the mother's cost. It is thus in the father's genome's best interest to make the mother invest more in his child than the mother would choose based solely on her own interests. The father doesn't talk the mother into doing more for the child; that message is encoded in the genome he passes on.

There is an apparent paradox at work here. If the father had a gene that made embryos suck up more resources, this gene would certainly do the trick now. But that same gene would also cause his grandchildren to consume more resources, re-gardless of whether they receive it from the man's son or daughter. Half of the time, the gene would be inherited pater-nally, which would improve its success in the society of genes. But half of the time, it would be inherited maternally, which would diminish its spread, as the embryo would consume the mother's resources at the cost of its future siblings. Over time, such a gene would not do well.

Thus, a gene cannot be successful if its only instructions are "keep sucking even if Mommy says it's been enough." Its message must include a clause that stipulates that the message to keep consuming is to be heeded when the gene is inherited from the father and ignored when it is inherited from the mother. Such a system is called imprinting. While your cells cannot normally distinguish between genes inherited from your father and mother, the chromosomal regions that hold imprinted genes are chemically modified to influence their expression. Some of the genes carried by sperm are imprinted to promote the growth of the embryo. To compensate for this, the mother responds by imprinting other genes in her egg cells that reduce the embryo's growth. In short, our genome reflects not only the arms race between our immune system and bacteria and viruses, but also the arms race between men and women, fought out between the two halves of our genomes.

Have you ever wondered why virtually all species that reproduce through sex contain roughly the same number of males and females? As the commercial breeding of animals illustrates, a single male is perfectly capable of fathering the children of very many females. So, in principle, humanity as a whole could have produced more children throughout its history if there were fewer men and proportionally more women. The answer is not simply that meiosis produces equal numbers of males and females by putting the father's X or its paired Y chromosome into each sperm with equal probability. After all, an equal number of males and females are also found in species such as the alligator, where sex is determined not by meiosis but by an altogether different mechanism: the temperature of incubation of the egg. When the alligator mother chooses a location for her nest, she determines the sex of her children: nests built on embankments are warmer and produce mostly males, while nests in wet marshes are cooler and produce mostly females.

Because each child's genome is half from the father and half from the mother, all the fathers in a parent generation will, as a group, have exactly as many children as the group of mothers do. In an animal breeding facility that has one bull and one hundred cows, with each calf inheriting half its genome from its mother cow and half from the one bull, members of the scarcer sex (in this case, the male) have more children on average. Putting it another way: if a society has more males than females, every additional woman is guaranteed to find a husband, while every additional man will be lucky to find a wife. This simple process will favor mechanisms that increase the birth rate of the scarcer sex, thereby restoring the 50:50 sex ratio.

As far as we know, mother alligators do not count the number of male and female alligator babies in their neighborhood before they decide on a location for their nest. But if the sex ratio in a population of alligators becomes distorted—say, for example, because a change in climate has caused more male than female alligators to be born—then any mutation that causes a shift in nest choice toward colder nests (and hence more female babies) would provide a fitness advantage, because female alligator babies will have an easier time finding partners later on. The three conditions of natural selection— variation, heritability, and fitness effects—would be met, and, with time, this mutation would become more frequent in the population. When the sex ratio again approached 50:50, the fitness advantage would cease to exist, and the mutation favoring the production of females would die out.

While the ratio of newborn human girls and boys is always close to 50:50, it varies between different human populations, an indication that there is heritable sex ratio variation among humans, too. When there is an excess of one sex in a population, natural selection will act to restore the equilibrium ratio of 50:50 by favoring the alleles that tend to produce the scarcer sex.

For cultural reasons, some societies value the birth of a boy more than that of a girl. If abortions are biased by the sex of the embryo, the ratio of male to female births is distorted. In China, due to sex-biased abortions, 120 boys are born for every 100 girls. Soon, there will be a surplus of 40 million young men. Given enough time though, natural selection will compensate, and the balance will be restored. But of course there is a more attractive solution to the problem. Moral considerations and extenuating circumstances aside, prospective parents that contemplate the abortion of a female embryo should decide whether they are more concerned with being cared for when they age (which, according to Chinese tradition, suggests a boy) or about having a grandchild (which strongly suggests a girl).

We have seen that sex evolved as an efficient, egalitarian mechanism that allows the society of genes to try out collaborations between the alleles of different genes. In this way, sex boosts the power of natural selection, helping the society to adapt to changed environments and to weed out deleterious mutations. Are all changes to the makeup of the society due to selection? Are there alleles that prosper due to chance alone?

4

The Clinton Paradox

Our true nationality is mankind.

—H. G. Wells

AS PRESIDENT, BILL CLINTON HAD BEEN AN ARDENT SUPPORTER of the human genome project—the quest to determine the exact letter sequence of humanity's genome. Starting in 1990, the project spanned thirteen years and was a roller-coaster ride of technological improvements, involving a surprising competition with a commercial company right before the finish line. Throughout, Clinton indulged the project in additional budget support. He was not disappointed. In speeches he gave after his presidency, he often described the amazing realization he had obtained through the human genome project for the bargain price of $2.6 billion.

In one of the Millennium Lectures in 1999, Eric Lander, a leader of the human genome project, told the audience at the White House that any two people on this planet are 99.9 percent identical in terms of their genomes. This struck Clinton as a fundamental insight. All wars, all cultural differences, all our destructive rivalries—all because of just a 0.1 percent difference between any two of us? Shouldn't this realization help us get over our differences and work together for the 99.9 percent that we share? The argument does have a seductive power: if all of us are 99.9 percent the same—why can't we all just get along? But as Lander also pointed out, there is another side to this argument. Recall that our genome is 6 billion letters long. While 0.1 percent sounds small, it amounts to a difference of 6 million letters between your genome and that of your neighbor. Would 6 million differences actually justify some rivalry?

You don't even have to go to your neighbor to find this level of dissimilarity. You yourself possess two copies of each chromosome, so you might as well compare the chromosomes you received from your mother with those received from your father. Your parents are about 99.9 percent the same, which leaves a 0.1 percent difference between the two sets of chromosomes you inherited from them. Does this mean that we should also be at odds with ourselves?

To understand what makes humans different from one another, we need to take a closer look at those 0.1 percent differences. You will recall that mutations are the result of the same sorts of accidental misspellings that often occur when we retype a document. The most common misspelling is a change of a single letter (or base) in the genome. Such single-letter differences are very common; the first estimate of differences reported to Clinton—the 0.1 percent—was based on these typos.

Another sort of typo is the insertion or deletion of a single letter or several letters. As research into the human genome continued, this typo turned out to be more common than originally thought. The number of copies of whole chromosomal regions, sometimes containing one or more genes in their entirety, can vary from person to person. That is, your neighbor's genome may have two copies of the *CCL3L1* gene (one on each of his two chromosomes 17), while there might be five in yours (two on your maternally inherited chromosome 17, and three on the paternal copy). If this were actually the case, you would be lucky: The *CCL3L1* gene produces a protein that blocks the entry points in your immune cells through which the HIV virus can enter. The more copies you have, the less likely you are to contract HIV.

The discovery of such widespread copy number variations pushed the percentage of differences up dramatically, to 0.5 percent, or 30 million letter differences between people. Would Clinton continue to argue that 30 million differences between people are too insignificant to explain why they are so often fighting one another? We call this the Clinton paradox: on the one hand, our genomes are 99.5 percent identical to one another, but, on the other hand, that difference of 30 million letters may not be negligible and is worth exploring in detail.

Height, skin color, and facial features are in large part heritable. Many of the more subtle variations that make you unique also have a component inscribed in your genes. Some of these variations predispose us differently toward diseases. For example, everyone has a set of genes that encode proteins, called hemoglobins, that are responsible for transporting oxygen through your body. A single letter mutation in one of these hemoglobin genes, if inherited from both parents, causes sickle-cell anemia. Interestingly, if your genome contained one defective and one normal copy of the gene, not only would you be fine but you would also be less prone to catching malaria. Such a genetic makeup would provide you with a significant fitness advantage in a region with a high incidence of malaria, and, accordingly, the mutated allele is relatively common in such regions. Individual mutations are often neither completely good nor completely bad; their consequences depend on circumstances, such as whether the allele is inherited from both parents and the conditions of the local environment. Mutations in the 20,000 genes in the human genome can set the stage for acquiring a disease. Thus far, more than 6,500 mutated genes have been linked to specific diseases. Most of these mutations do not guarantee that the disease will develop; if they did, they would quickly be removed from the society of genes through natural selection. Instead, due to complex interactions with the environment and the other alleles in the genome, they represent nothing more than slight inclinations toward disease. A complex series of steps has to occur before the disease can emerge, just as in cancer: one mutation on its own is often not enough to lead to a disease.

In and Out of Africa

As genome sequencing technologies continue to improve, it is no longer prohibitively expensive to have your genome sequenced. But reading your genome cover to cover would not be enlightening. It would be more fruitful to compare your genome's sequence of letters to that of another person and look for the differences. It will be hard to appreciate the significance of each specific difference, but their sheer number can provide valuable information. From Clinton's paradox, you know to expect about 30 million different letters (including the deleted or duplicated parts). If you compared your genome first to your sibling's, then to a cousin's, and then to that of a stranger, you would find that the number of differences increases. This does not come as a surprise: we expect to be more similar to close relatives than we are to strangers. The more similar two genomes are, the more recently they had a common ancestor; in other words, the more closely they are related to each other.

Take the genomes of your parents, your grandparents, and your great-grandparents, plus your own. You inherited half of your genome from each of your two parents, one quarter from each of your four grandparents, and one eighth from each of your eight great-grandparents. This means that one-quarter of your genome is identical to the corresponding quarter of your maternal grandfather's genome (ignoring the very small number of newly introduced mutations). The remaining three-quarters of your grandfather's and your genome will differ by 0.5 percent of their letters, the same as for two unrelated people. Thus, overall, your genome differs from that of your grandfather's by 0.375 percent—one-quarter less than 0.5 percent. Similarly, in comparison to unrelated people, you are one-half less different from your parents, and one-eighth less different from your great-grandparents.

Imagine building your family tree from these genomic similarities by placing photographs on a table, each representing one person in the family. Connect with a string the two people with the most similar genomes. Then connect the two people with the second most similar genomes, and so on. Continue this process until you get a tree that connects everybody with everybody else. You will have connected everybody with his or her two parents, making a family tree based on your genomes. This method becomes more involved

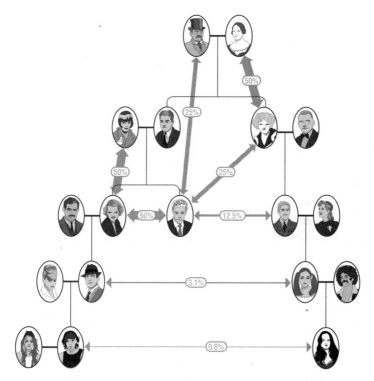

Figure 4.1: A genomic family tree. The percentages indicate the degree of similarity of each generation's genomes over and above what you would expect between strangers.

when you include siblings, as siblings share the same fraction of their genomes as do parents and children—half. To figure out the precise location of siblings on the family tree would require a more detailed look at their actual genomic letter sequences.

Within each set of siblings, the genomes would be ½ identical (Figure 4.1): each got a random ½ from father and mother, so for any gene, the probability that the two siblings both inherited a given allele from their mother is $\frac{1}{2} \times \frac{1}{2} = \frac{1}{4}$. The probability that they both inherited the father's allele is equally $\frac{1}{4}$. Together, the probability that they either both have the mother's or the father's allele is $\frac{1}{4} + \frac{1}{4} = \frac{1}{2}$. For first cousins, the commonly inherited fraction can be calculated as $\frac{1}{2}$ (if, say, the mothers of the two cousins are sisters), $\times \frac{1}{2}$ (the child of the firstborn sister shares $\frac{1}{2}$ of the genome with her mother), $\times \frac{1}{2}$ (and so does the child of the second born). Thus, first cousins share $\frac{1}{8}$ of their genomes. This does not mean that cousins are only $\frac{1}{8}$ identical in terms of the letters in their genomes. Recall that any two strangers are 99.5 percent identical. What this means for a pair of cousins is that they don't differ by 0.5 percent, as two strangers would, but by 0.4375 percent, $\frac{1}{8}$ less. As generations go by, the degree of similarity among family members keeps diminishing. However, as members of the greater human family, we are always still 99.5 percent alike.

What would the family tree of all of humanity look like? As you went back in time, you would discover more and more distant relatives of yourself. One generation ago, you had two parents; two generations ago, you had four grandparents, then eight great-grandparents, then sixteen great-great-grandparents, and so on. If you continued to follow that logic, then forty generations back you would have had a trillion great-great- . . . -great-grandparents. That's 200 times the number of people living on the planet today, a ridiculously high number of alleged great-great- . . . -great-grandparents. This is because when you go far enough back in time, the people on your mother's side and the people on your father's side were often the very same people. If, for example, your grandparents had been first cousins, then you should count their shared grandparents (two of your great-great-grandparents) only once. The history of humanity is a web of relationships, with family lines splitting and merging again. And that tangled family tree tells fascinating tales of our ancestry.

The principles of genomic relatedness make it possible to construct genomic family trees that connect people all over the world (Figure 4.2). Such a map can be drawn at different levels of detail, but for our purposes let's look at a tree based only on the strongest signals of relatedness. Certain relationships in the tree are not surprising. For example, the genomes of French people are closely related to one another, and they are also closely related to the genomes of Belgians, Swiss, and Germans, whose countries border France. Genomes from these four countries are fairly similar to genomes in other European countries. It turns out that in general, genomes within most of the continents are, broadly speaking, quite closely related to each other.

The relationships revealed by this tree do much more than merely reflect the present-day distribution of people around the world. Whenever an individual or a group of people moves to a new region, they are, after all, the vehicles of their genomes. They retain their genomic similarity to their place of origin, slowly diluted through the occurrence of new mutations. The similarities among our genomes therefore make it possible to reconstruct the history of early human migrations.

There is one major exception to the rule that people from the same continent are generally quite closely related. If you compare someone from one African population to someone from a different African population, they are likely to be genetically

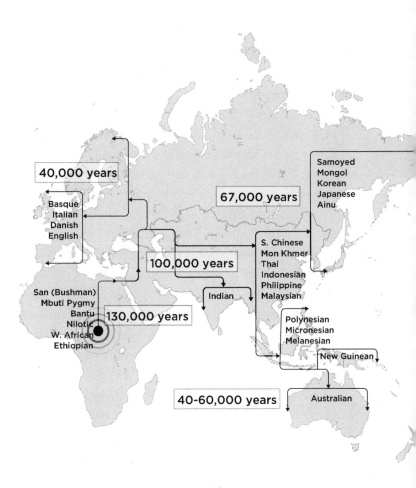

Figure 4.2: The family tree of genomic relationships worldwide, reflecting the migrations of humanity from Africa to the other continents. The numbers give an indication of how long ago these migrations happened. After the first humans ventured out of Africa almost 100,000 years ago, it took more than 80,000 years before they finally reached South America, 13,000 years ago.

Eskimo
Chukchi

20,000 years

N. Amerind

C. Amerind

S. Amerind

13,000 years

more dissimilar than a Korean is from a German or than an Alaskan is from an Australian aborigine. We have to go far back in time to understand why this is the case. The pattern of similarities in our genomes indicates that anatomically modern humans evolved in Africa roughly 400,000 years ago. Different groups in Africa lived in isolation for long enough to diverge into populations that can be distinguished by their genomes. Then, less than 100,000 years ago, a small group migrated north, crossing the Sahara into the Middle East. Compared to the people that stayed back in Africa, the migrators were a relatively homogeneous group. We know that the group was made up of a few extended families, because the alleles they took with them are just a minor fraction of the alleles still present in Africans today. This odyssey was astonishingly successful: their descendants found new homes throughout the world.

The genomic record reveals the steps that humans took as they expanded their domain. From all the genomes found outside Africa, those in the Middle East, the first way station of the migration that later settled the rest of the world, are most closely related to African genomes south of the Sahara. From the Middle East, some of our ancestors traveled farther east along the coast, settling East Asia and Australia. A bit later, other groups headed northward out of the Middle East and ended up in Europe. As recently as 20,000 years ago, people from Asia crossed over Alaska into North America, taking almost another 10,000 years to spread into South America. About 4,000 years ago, Oceania, the area that includes the Pacific islands, became the last region in the world to be settled by humans.

Because of these migrations, everyone outside Africa (with the exception of the descendants of Africans who left that continent within the past few hundred years) is descended from a few small groups of hunter-gatherers that crossed the Sahara. The genomes of those who remained in Africa retained their original differences, which is why the biggest genomic disparities are found there. But let's remind ourselves that every human genome remains nearly identical.

Bacteria provide further genomic evidence that the human species had its beginnings in Africa and migrated out from there. When the first groups of humans left Africa to populate the world, they were not alone. Another species traveled comfortably in their stomachs: *Helicobacter pylori*, or *H. pylori* for short. At least half of all present-day humans are infected with this bacterium, making *H. pylori* the most widespread pathogen. There are no long-term consequences for the vast majority of infected people, but in some people infection can develop into gastritis, the acute or chronic inflammation of the stomach. Over a lifetime, a person with an *H. pylori* infection carries a 10 percent greater risk of developing chronic ulcers and a 1 percent greater risk of evolving stomach cancer than those without the infection. *H. pylori* lives exclusively in the human stomach. Children acquire the infection from the people around them, so *H. pylori* largely remains in the families of its carriers.

When comparing the genomes of *H. pylori* isolated from stomachs of people living in different geographical regions, we can reconstruct the "migration" history of this bacterium. As expected from its close association with humans, what we see closely resembles what we see for human genomes. The genetic diversity of *H. pylori* decreases with increasing distance from Africa, just as it does for human genomes: all *H. pylori* genomes outside Africa are more similar to one another than are *H. pylori* genomes from different African regions. Mirroring the movements of human migrations, *H. pylori* first crossed from sub-Saharan Africa into the Middle East, then moved on to Europe and Asia; from Asia, it continued into Australia, the Americas, and, finally, Oceania.

Moving closer in time, *H. pylori* variants typical for members of the Bantu population spread through Africa via a vast migration that started out from the Bantu homeland in northern Africa about 4,000 years ago and reached southern Africa 2,700 years later. We can follow the colonial expansion of *H. pylori*, conquering the world in European stomachs from the 1500s on, now detectable as typical European *H. pylori* found at low frequencies in the stomachs of Native Americans, Africans, and Australians. Traces of West African *H. pylori* have been found in the stomachs of some Native Americans, a consequence of the slave trade, which began in the seventeenth century and ended in the mid-nineteenth, a mere instant ago in evolutionary history.

Evolution You Can Taste and See

Most of the millions of letters that differ among some human individuals contain little information from which to reconstruct human history. For 85 percent of the positions in the genome that occur in two different versions—for example, some people may have a C where others have a T—you are equally likely to find different letters when you compare one of your chromosomes to your next-door neighbor's or to a chromosome from the other side of the world. Or, for that matter, to the matching chromosome inside your own genome; remember that the parts of your genome inherited from mother and father are different, too.

In other words, the vast majority of genetic variation does not distinguish different ethnic groups. Clinton then did have some reason to take comfort in a brotherhood and sisterhood of humankind: only about 15 percent of the differences found among human genomes can help distinguish people from different populations—that is, groups whose members rarely married outside their own circle throughout their recent evolution. And very few alleles are specific to a particular population. For that, all members of one population would have to have the same letter at a certain position and everybody else on the planet would have to have another letter at that position. What is the evolutionary reason behind the existence of these few population-specific alleles?

Most alleles that are specific to certain populations are related to the environment. A prime example is skin color, an important adaptation to geographic regions. Skin color is the result of a compromise. Darker skin provides protection from the sun's ultraviolet (UV) rays, which is particularly important in regions close to the equator. If too much UV radiation penetrates the skin, it can, over time, damage DNA and accelerate the evolution of skin cancer. This is why people with light skin should use sunscreen. However, absorbing too little UV is also damaging. Our bodies use UV radiation to make vitamin D, an important molecule that enables us to absorb vital chemicals, such as calcium and phosphate, in our intestines. Without enough UV radiation filtering through the outer layers of our skin, there is not enough vitamin D in our bodies, and without that, we are at a risk of developing soft bones—a condition called "rickets" in children.

The best compromise between protection from skin cancer and enough synthesis of vitamin D is to set the skin's pigmentation to a particular shade that lets through just enough UV to make the required amount of vitamin D, but not more. The "setting" has been done by natural selection: according to the local sun intensity, skin color genes that caused shades too dark and those that caused shades too light were outcompeted by alleles providing the best balance. Close to the equator, strong radiation selects for strong UV protection and a correspondingly dark skin. But at latitudes greater than 30°, the sun's rays pass the atmosphere at a much lower angle and are much weaker. This severely compromises the production of vitamin D in dark-skinned individuals. The optimal skin color can be accurately predicted by a simple formula. In agreement with that prediction, different versions for skin color genes predominate the society of genes in different geographic regions (Figure 4.3).

Because natural selection is a slow process, your skin color may not reflect the level of UV radiation of your current home. What the color does indicate is the UV radiation experienced by many generations of your ancestors. The mobility of humans in today's globalized world is the major reason many lightly colored humans need sunscreen, though the increase in UV levels in recent times has further intensified the need for protection. Conversely, people with dark skin who live at higher latitudes may often profit from vitamin D supplements in their diet.

Skin color is, of course, not fixed. Through tanning, we can also "adapt" to solar radiation levels in real time. Skin color is determined by the activity of the pigment melanin, which is produced in specialized cells in the bottom skin layer. Melanin absorbs light, thereby protecting the cells in deeper layers. When the exposure to too much UV light leads to DNA damage in your skin, more melanin is produced. However, this ability of on-demand melanin production is limited, which is why we are already born with a basal skin color that reflects the intensity of solar radiation experienced by our ancestors.

The ability to digest milk is another example of a genomic variation that distinguishes groups of people. Because, as mammals, we naturally feed on breast milk as babies, our genomes encode a mechanism for digesting lactose, the primary sugar in milk. One of our genes encodes the protein lactase, a molecular machine that cuts lactose into two smaller sugars, glucose and galactose. For most of human history, early childhood was the only time of life when milk was part of the diet, and the lactase gene originally was programmed to turn itself off once a child stopped breastfeeding. The hunter-gatherer diet was plant-based, supplemented with some meat or fish. For many millennia, it thus made perfect sense to conserve bodily resources by switching off lactase production once suckling ceased.

An enormous dietary change began around 8,000 BC, when farmers in the Middle East started to domesticate animals and milk them. Today, 10,000 years later, 90 percent of people in Western countries are lactose tolerant, meaning they

UV Radiation

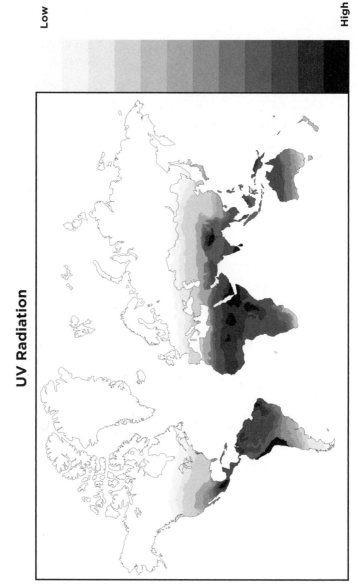

Skin Color

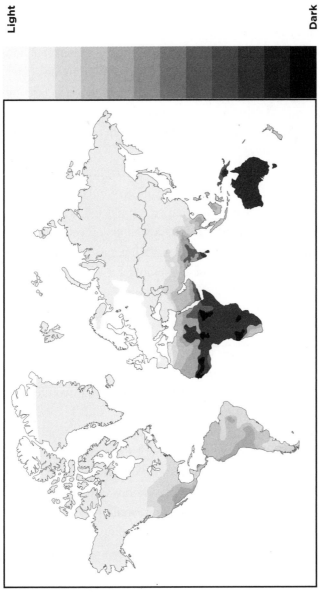

Light

Dark

Figure 4.3: Levels of ultraviolet (UV) radiation and skin color of native people around the globe. The alleles for a locally optimal shade of skin color dominate the societies of genes throughout much of the globe. Exceptions are Central and South America: these regions were settled less than 20,000 years ago by people with light skin, likely providing too little time for natural selection to reach the optimal darker color.

are able to digest milk as adults. Their regional societies of genes evolved to retain expression of the lactase gene long after weaning. But in some Asian and African populations that have no tradition of using dairy cows, no such evolution took place, and only about 10 percent of these people produce lactase into adulthood. The highest level of adult lactose intolerance is found in Native Americans, who came into contact with dairy farming only in recent centuries.

To remove the switch that turns the lactase gene off after infancy, only a single letter in the gene's control elements needs to be replaced. Lactose intolerance is rarely felt before the age of six, somewhat later than the typical age of weaning in traditional hunter-gatherer civilizations. It is easy to imagine that, in an early, naturally lactose-intolerant tribe with domesticated cattle, a girl born with a chance mutation to the lactase switch had a big advantage. By retaining lactase tolerance beyond age six, this child had access to a valuable food source, which meant greater chances of survival in times of food shortages and a lower risk of suffering from conditions linked to nutrient deficiencies. Because of these advantages, she would likely have had more children than other women did. The new mutation would be located on only one of her chromosomes, and so half of her children would inherit the mutation, allowing them to enjoy the same advantages, including having a high number of children. The three conditions of natural selection would be met. Slowly, the mutation would have replaced the lactose-intolerant allele of that cattle-herding tribe.

On the evolutionary timescale, the domestication of cattle 10,000 years ago is a very recent event. We now have compelling evidence that lactose tolerance emerged in Europe in just the past 3,000 or 4,000 years, thanks to DNA extracted from 3,800–6,000-year-old European skeletons and the 5,400-year-old ice mummy known as Ötzi, discovered in the Tyrolean Alps in 1991. None of these specimens had the lactose tolerance mutation in their DNA, indicating its rarity in the ancient society of genes.

It is somewhat ironic that today, lactose intolerance is considered a deficiency; in fact, it was the natural state of health for most of human history. If you are lactose intolerant, you simply have an allele that is slowly going out of style in the society of genes. It could have gone another way, though. The Maasai of southern Kenya and northern Tanzania have a long tradition of milking cattle, yet are mostly lactose intolerant. They curdle their milk, which reduces its lactose content, and this might have mitigated the advantage of adulthood lactose tolerance.

The Lucky Gene

The genetic variants underlying different skin colors or lactose tolerance are of precisely the kind that Clinton was concerned with—those that distinguish people in noticeable ways. The evolution of these traits is a clear demonstration of the power of natural selection. But they turn out to be the exceptions. The vast majority of the 30 million differences between each of us are not the results of adaptations to our different environments. Why are they there, and what role do they play?

Most of the 30 million differences between you and your neighbor are of no consequence to either of you. The genes that cooperate with one another to build and control your body for their common good are spaced out along your chromosomes, interrupted by long stretches of DNA that do not contribute to this endeavor. Many of the 30 million differences fall in these islands between the genes.

Another reason the differences seldom matter is that—as we saw in Chapter 2—the "useful" parts of your genome are written in a rather forgiving code, allowing them to be read correctly despite typos. The English language is not unlike this: comsirer how eahily yhu caw unjerstanf thhs tessed-ud sentccce. What is more, the genome does not have a defined "space bar"; regions that separate the important bits can have an arbitrary sequence of letters. And, finally, many of the differences among people are repetitions of regions already present elsewhere in the genome.

If a mutation in your genome has no functional significance, why doesn't it simply disappear? Natural selection requires a fitness effect of the variation, so why would a mutation without such an effect—a neutral mutation—ever become widespread? As it turns out, its persistence in the genome may be caused by nothing more than chance.

The influence of chance on genome evolution can be illustrated by an experiment with fruit flies. To do this experiment, you would first select one hundred fruit flies according to the following two criteria: half of them must be male, and half must be female; and half must have white eyes, and half must have red eyes (red being the normal color of fruit fly eyes). Eye color is determined by a single gene and is of no consequence for the flies: white-eyed flies see just as well and are neither more nor less attractive to the opposite sex. You would

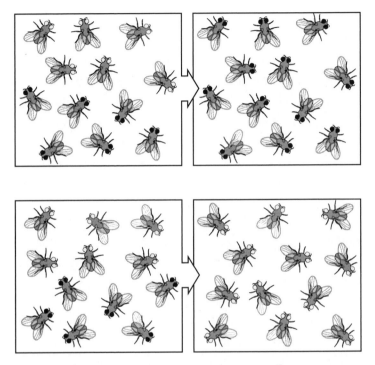

Figure 4.4: The fruit fly experiment. In an experiment shown by the top row, no white-eyed flies remain in the population after several generations. In another experiment shown by the bottom row, the reverse has occurred.

then place all the selected flies together in a comfortable closed container (Figure 4.4).

After one generation, you might find that the fraction of flies with white eyes has increased slightly, from 50 percent to 55 percent, simply because of the randomness involved in sexual reproduction: even if all flies are equally fit, some will have more children than others. In the second generation, the white-eyed fraction might go, again by chance, back down to 52 percent, and then back up to 56 percent in the third generation. The mutation for white eyes can be said to be drifting in the society of genes. After many rounds of reproduction, all flies in the population might have white eyes, again, just by chance. At that point, when there is no longer any variation in eye color, the drifting stops. The reverse could just as easily happen. Since white-eyed flies do not, on average, have more children than red-eyed flies, the white-eyed mutation could have been the one to die out. And whichever color dies out, that variation is lost, gone forever—unless, of course, an eye-color mutation were to occur again.

The same rules of chance apply to a new mutation on a human chromosome. The only aspect of an allele that is relevant to natural selection is how efficient it is in generating copies of itself. If this ability is not altered by the new mutation, then the fate of that mutation is not influenced by natural selection; it is subject only to chance. Because the new mutation starts life as a minority—a lone allele on a single chromosome amid all the nonmutated alleles in the society of genes—it will most likely die out within a few generations. Such is the fate of most new mutations. It would be as if you started the fly experiment with 99 red-eyed flies and one white-eyed fly. The odds that white eyes would eventually win out over red eyes would be very low. However, it is still possible that by chance the mutation could rise in frequency, eventually replacing the once dominant allele.

What the fruit fly experiment shows is that, eventually, one allele will win out. Having two functionally equivalent alleles in the society of genes is rarely a stable situation. When we see a particular variation in the population, what we are actually looking at is a freeze-frame of a moment in evolution

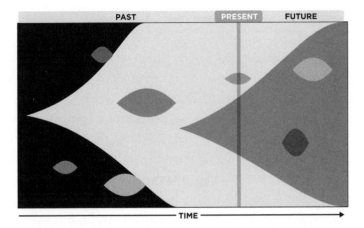

Figure 4.5: The rise and fall of new alleles for one gene. The arrow indicates the direction of time. Each vertical slice of time shows the distribution of alleles at a given moment. Initially, 100 percent of the bacteria have the black allele. The leftmost tip of each distinctly colored area indicates the origin of a new allele through a mutation. Most alleles are quickly lost again, but sometimes a new allele rises to dominance in the society of genes, pushing its ancestor to the wayside.

(Figure 4.5). The fate of the allele has not yet been sealed, but in time it will either become extinct or take over.

One crucial thing to know about each allele is how popular it is in the society of genes—that is, the fraction of genomes that carry this version. For example, the allele related to lactose tolerance occurs at a frequency of about 50 percent of all human genomes combined. How popular are the 30 million typically? We find that only very few occur at appreciable frequencies, as expected if they provided fitness advantages similar to the lactose-tolerance mutation. Conversely, most of the 30 million alleles are rare, occurring in just a few percent of genomes. Most of the variations found in the society of genes are nothing but random fluctuations in the game of life.

Africa's Genomic Cornucopia

Studies of human genetics often aim to include genetically diverse individuals, comparing, for example, DNA sequences from people of European, Chinese, and African descent. As this chapter has shown, if you instead compared genomes from a selection of different African populations, these would cover much more diversity than one could ever get from comparing Asian, European, American, and Australian genomes. Moreover, adding genomic varieties from outside Africa would not increase diversity much over the variety found in Africa alone: nearly all variants observed in individuals outside Africa are also present in African genomes. Not so vice versa: many genetic variants are found in Africa alone.

Much of this within-Africa genomic variation does not affect the appearance or the fitness of its carriers, but a small fraction does. Talents, like physical appearances, are to some extent heritable, passed down with our ancestors' genomes. If a certain talent—say, the ability to sprint fast—is linked to specific genomic variants, it is likely that there will be a population somewhere in Africa where many individuals have that particular variation, both men and women; much more likely, in fact, than the probability of that variant being frequent among, say, European genomes. This follows simply from there being overall more genetic variability inside Africa than outside. If another variation makes you run exceptionally fast over long distances, again it is much more likely that this variant will be frequent somewhere in Africa than, say, in Asia. It does not depend on what that ability is: just because there is so much more variation among Africans than there is among people elsewhere, it is likely that the people who are genetically best equipped to be good at something live somewhere on the African continent. This does not mean that all Africans are more talented in any specific discipline; on the contrary, more variation means that we may well also find the slowest sprinters somewhere in Africa.

In recent decades, we have seen the consequences of this uneven distribution of diversity across the world in action in many athletic disciplines at the summer Olympics, which have been dominated by men and women of African descent. These days, a Chinese or French sprinter in the 100 meter finals would be a notable exception, unless that athlete had recent African roots.

It has not always been like this. Circumstances play a major role in whether a genetically inscribed talent is expressed. Jesse Owens, an African American, competed, along with seventeen other African American athletes, in the 1936 Olympics in Nazi Germany. No native African athletes competed. South Africa participated, but sent only white athletes to the games. To the chagrin of the Nazis, Owens collected gold medals for the long jump, the 100 meter, the 200 meter, and the 4×100 meter relay. Nine of the other African Americans on the U.S. Olympic team also captured medals. The same genomically inscribed talents we observe in the Olympics' finalists today were of course already found among Africans in the 1930s. But well-tuned genomes are not enough to excel at a given task. You also need appropriate training and support, which was not accessible to native Africans still dominated by colonial powers.

Athletes of European or Asian descent still dominate many sports that are technically challenging or expensive—likely not because they are genetically better endowed for these sports, but because the means or the motivation for practicing them are not widespread in Africa. But it seems not entirely unlikely that one day the children of a population somewhere in Africa will start playing chess, and a few generations later a succession of African champions will dominate the international chess scene.

In Spite of Our Genes

Go to a comedy club in New York City on any night, and many of the jokes you'll hear have to do with racial slurs about the differences between people: African Americans, Mexicans, Asians, Arabs, Jews. Are we any wiser after our worldwide genomic survey to say whether there is a genetic basis for this racism? To be sure, there are differences between the people of the world, and many of these differences are inscribed in our genomes. But as Clinton would point out, these differences seem hardly big enough to justify any discrimination. So why does racism persist?

For an informed discussion, it is important to see how easily discrimination arises in a society of genes. Consider the green beard effect. Imagine a mutation that produces an allele with two consequences: people who inherit the mutant allele grow a green beard, and they help out people who also have green beards. As long as the help is more valuable for the recipient than it is costly for the benefactor—a reasonable assumption in many situations—this behavior would result in an increased fitness of the green beard allele: it gets more than it pays, even if cost and benefit occur in different individuals. We can, of course, replace the green beard by any other conspicuous sign caused by a specific allele.

The green beard theory was developed by W. D. Hamilton, one of the great theoretical biologists of the twentieth century (the name of the theory was coined by Richard Dawkins, who helped popularize the concept). Hamilton examined the evolution of social behavior. Generalizing the green beard idea, he argued that altruism—behavior that is costly to you but benefits others—pays off for your genes if those receiving the benefits are more closely related to you than the average person in your population. That is why we tend to support our children, our siblings, and our cousins.

The flip side of this insight is that spite—behavior that is detrimental to others without giving you any direct benefits—pays off for your genes if those suffering the consequences are less closely related to you than average. This is because spite puts alleles that are mostly relatively distant from yours at a disadvantage, thereby giving your closer relatives an advantage in comparison; again, you are, on average, furthering the cause of your own alleles. This is the general theoretic basis of racism: it pays for your alleles to treat badly those that are less likely to carry the same alleles.

While green beard genes have been found in ants, slime molds, and fungi, no "racism gene" has yet been identified in humans. That racial discrimination has been widespread throughout recorded history suggests that it is there for a reason, and this reason is quite possibly the favoring of variants of the green beard type by natural selection. It is an interesting thought that such variants don't even have to be genetic but can also be cultural. The same logic of natural selection that applies to genetic variation also applies to cultural variation: if there is cultural variation that affects the number of offspring, and if children inherit the culture of their parents, then the "fitter" variant will increase in frequency.

To see how this mechanism could work, consider a simplistic example. Imagine a population split into two equally sized groups of people. One group, the egalitarians, firmly believes in extending help to all individuals regardless of their background. The other group, the elitists, has a culture of strongly favoring its own members, identifiable by their hairstyle. The elitists will receive twice as much support than the egalitarians, as they get help from the members of both groups. If, as a consequence, they raise more healthy children who then also become elitists, it is inevitable that the egalitarian idea will lose out over the generations—even if it may be superior on philosophical grounds.

In this sense, Clinton was wrong: even if we are more than 99 percent identical, both theory and history suggest that a small number of selfish genes (or even selfish ideas) are enough to underpin racist behavior. And not just in humans. When badgers are infected with tuberculosis, they emigrate from their own group (those closely related to them) to a neighboring group (those less related to them), thereby infecting the "others."

What has the potential to distinguish us from badgers is that we can prove to be more than the slaves of our genes. We have the choice of placing ideals over alleles, and to be more than the sum of our genes.

We find that many alleles do not have a selective advantage over their competitors, and thus their fate is in the hands of chance. Is it simple to predict the functional consequences of individual alleles? Or do an allele's consequences depend on who inherits it?

5

Promiscuous Genes in
a Complex Society

Things derive their being and nature by
mutual dependence and are nothing in themselves.

—Nagarjuna

IN HIS 1929 SHORT STORY, FRIGYES KARINTHY, A HUNGARIAN
writer, proposed the notion that any person can be connected
to any other person on the planet through a chain of acquain-
tances that has no more than five intermediaries. The theory
was popularized by John Guare's successful 1990 play *Six
Degrees of Separation* and its 1993 film version, which, in turn,
inspired the game Six Degrees of Kevin Bacon. In this game,
a player is given the name of an actor and tries to connect that
actor with Kevin Bacon. For example, if the actor is Harrison
Ford, the player would get a Bacon score of two by making
these connections: Ford appeared in the 2008 movie *Indiana
Jones and the Kingdom of the Crystal Skull* with Karen Allen,
and Karen Allen appeared with Kevin Bacon in the movie
Animal House. Very often, surprisingly few links are needed,
since Kevin Bacon played in many diverse movies. The game
would work nearly equally with other actors, though, since
actors tend to play in many movies over the course of their
careers, and each movie typically includes many other actors.
This set of relationships is a network that keeps expanding
outward (Figure 5.1).

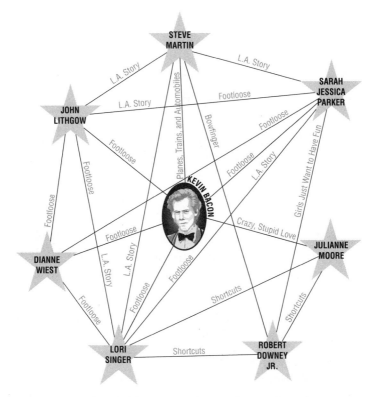

Figure 5.1: An expanding network of actors, linked by movies in which they co-starred. Every actor in the figure can be linked to Kevin Bacon in at most two steps.

Peas, Brother

Gregor Mendel (1822–1884) is known today as the father of genetics, but his work was not appreciated until 1900. Coming from a poor farming family, he was able to attend university by becoming an Augustinian monk. After university, he lived in an Austrian monastery, where he worked on his ground-breaking experiments on heredity. The bishop responsible for

Mendel's monastery did not approve of his attempt to study heredity in mice because it involved sexual reproduction, so Mendel changed his subject to peas—pleased that "the bishop did not understand that plants also have sex." Following a series of brilliantly executed experiments and analyses, Mendel published the results of his years of experiments in the well-respected *Proceedings of the Natural History Society of Brünn*, even though the significance of his work was not recognized for many years. To begin to explain how genes work with other genes, let's look at one of Mendel's experiments.

Mendel selected several traits to study, with seed color, pod color, stem length, and flower color among them. He would, for example, cross-pollinate a female plant with green seeds with a male plant with yellow seeds and follow their offspring through several generations. He found that each trait behaved as a coherent and indivisible unit—seeds were either green or yellow, flowers were either white or purple. When he crossed a male plant with yellow seeds with a female plant with green seeds, 75 percent of the peas in the second generation were yellow and 25 percent were green (Figure 5.2). From this simple ratio, he inferred that a plant must have two copies (what we now call alleles) of an entity that determines color and that comes in two types (one leading to yellow and one to green seeds). He also learned that the yellow version of the seed-color allele is dominant: that is, if a plant has one copy each for yellow and green, it will still produce yellow seeds. Green seeds will be produced only if a plant has two copies of the green allele. Out of the four possible combinations (YY, YG, GY, and GG, where Y and G are the yellow and green alleles), three have the Y allele, which is dominant, and so

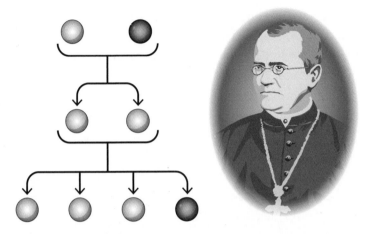

Figure 5.2: Gregor Mendel, father of genetics, and his experiment on peas. In cross-pollinating plants that either produce yellow (light gray) or green (dark gray) pea seeds exclusively, Mendel found that only yellow seeds are produced in the first generation. However, the following generation produces a 3:1 ratio of yellow to green seeds.

will produce yellow seeds. Mendel's breakthrough discovery marked the beginning of the science of genetics.

Guilt by Association

Mendel found that, in cross-pollinated plants, heritable traits were never a blend of the two parents, as was thought to be the case at the time. For example, seeds were either yellow or green; no intermediate colors were produced. This seemed to indicate that there is a simple one-to-one mapping between genes and traits, theoretically enabling biologists to focus on any single trait and find the gene and alleles that caused it. In this formulation, there would be one gene for every heritable feature: one for a bump on your nose, one for the color of your hair, and another one for the length of your index finger. Similarly, for every heritable disease, one causal mutation could be singled out.

Reality turned out to be much more complex. We have learned that we cannot assign most heritable diseases to mutations in a specific gene, even though for a sizeable minority of diseases we can. For example, one of the authors of this book has Milroy's disease, which is caused by a single letter change in the *FLT4* gene, a gene involved in regulating the development and maintenance of the lymphatic system. This single mutation leads to the replacement of one amino acid in the protein encoded by *FLT4*, which makes the protein defective, causing problems in the lymphatic system. Milroy's is a clear example of what we call a "Mendelian disease": a disease caused by just one defective gene.

Typically, though, the link between diseases and genes is not one-to-one. Parkinson's disease—a progressive disorder of the nervous system—is caused by mutations to any one of three genes. In the normal course of events, the three genes work as a team to decompose certain proteins so that they will not accumulate in brain cells. If even one of these genes loses its ability to function, the decomposition fails, the proteins build up, and brain cells malfunction as a result.

Then there are situations in which mutations in a group of genes that work together can cause different versions of a disease. In Chapter 4 we looked at the evolution of lactase, which breaks down lactose, the main sugar in milk, into glucose and galactose. Galactosemia is the inability to metabolize galactose. If an infant with galactosemia is given milk, toxic and often deadly levels of substances made from galactose build up in the infant's system and damage the liver, brain, kidneys, and eyes. Normally, a chain of three chemical reactions converts galactose into digestible glucose, and the three reactions are each governed by a different gene. A mutation in any one of these genes will result in a variant of galactosemia. A mutation in the gene responsible for the first of the chemical reactions causes clouded eyesight, a relatively minor form of the disease. A mutation in the second gene causes classic galactosemia, which, if untreated, leads to developmental problems and liver disease. Mutations in the third gene can lead to either a severe or a mild form of the disease. Because the symptoms of all three variants are similar, the resulting diseases are still subsumed under the same medical term, galactosemia.

All of the above diseases can be traced back to the malfunctioning of individual alleles. This sounds hopeful: theoretically, it should be possible to systematically find out what each gene is responsible for, and which disease occurs when it malfunctions. Once doctors have a corresponding mapping in hand, they might examine your genome and prescribe appropriate medication—even before you have any symptoms.

Let's take a look at how we link a mutation to a disease. Say you are planning to investigate Crohn's disease, a severe inflammation of the bowels. First, you would need to assemble a fairly large group of subjects—let's say one hundred people— half of whom have the disease and half of whom do not. Next, you would determine each subject's genome sequence, a task that is becoming increasingly less laborious and expensive. Then you would go through the genomes, one DNA position at a time, to see whether the letters the subjects

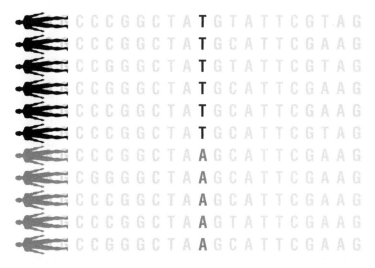

Figure 5.3: Genome-wide association studies aim to associate particular alleles with medical conditions. The rows list corresponding stretches of DNA from different individuals. The dark individuals in the top six rows are patients afflicted by a disease. The bottom five rows are from healthy individuals. The darker DNA letters in the center column differentiate perfectly between the two groups: all the patients have a T at the genomic position, and all the healthy people have an A. This is a clue that the allele with the T mutation may be responsible for the disease.

have at that position are correlated with their states of health. Hypothetically, you might find one allele—say a T in position 5,727,514 on chromosome 16—that is present only in the fifty subjects with Crohn's disease (Figure 5.3). Since a perfect 50:0 pattern is highly unlikely to occur solely by chance, you could reasonably conclude that this allele is a very good predictor for Crohn's disease. Correlations as strong as 50:0 are extremely rare, but if the subject group is large enough, important alleles can still be detected. This type of investigation is called a "genome-wide *association* study" (or GWAS for short).

When a GWAS was carried out for Crohn's disease on 6,333 affected individuals and 15,056 unaffected ones, researchers found seventy-one regions of the genome that influence the odds of developing the disease. Each region is termed a "risk locus," because particular alleles in each region lead to an increased likelihood of developing the disease. Surprisingly, however, the GWAS found that only about 25 percent of the Crohn's sufferers had the disease-associated alleles in any of the seventy-one regions. Thus, many people with the hereditary form of Crohn's disease do not have any of the identified alleles. Perhaps disturbances in many more than the genes found in the seventy-one regions are often involved; a GWAS on a much larger scale would be needed to confirm this idea.

There is also another possibility. Only alleles that are compatible with each other can successfully work together in one genome—there are society members whose protein products just do not interact well. It is likely that at least for some patients, the disease occurred because variants of two of the regions came together for the first time in their genomes and did not get along. These same alleles might have worked perfectly well in other combinations in the mothers and fathers of the patients, but they led to a disease state when combined. Such interactions between alleles are called epistasis. We saw epistasis at work in Parkinson's disease, which results from a failed interaction among three genes. Parkinson's represented a simple case: usually, as in Crohn's disease, many more genes are involved.

Now that hundreds of GWASs have been carried out for a host of diseases, we can say that most diseases are influenced by a large number of genes. In addition, because most variants associated with a given disease also are present in some healthy people, allele interactions probably do play a big role, as do interactions between the alleles and the environment. Even for those diseases that have been researched for years, each new GWAS reveals several more genes and interactions not previously implicated. Thus, the study of heritable diseases emphasizes that to perform a given function, many different genes have to come together in coordinated and complex patterns.

The Rotting Ship of Theseus

Your body is a very complex machine, and most processes it runs are too involved to be performed by the protein produced from a single gene. For example, to convert the sugars in the food you eat into usable energy, dozens of independent chemical reactions must take place. Each of these reactions is organized by a different enzyme, specific proteins that accelerate (catalyze) reactions that would otherwise proceed too slowly. The work of each enzyme is highly dependent on the proper functioning of all others involved in the same process: if one of the preceding steps fails, the enzyme would be out of work; if one of the following steps malfunctions, the enzyme's products would accumulate, often with disastrous consequences.

Let's consider another example from Chapter 4, where we explained how your skin's color evolved to be optimal for the region of the world where your genes spent most of their recent millennia. Skin color may seem like a simple feature, but alleles in at least fifteen different genes affect it. As we saw in

Chapter 4, until humans began to leave Africa 100,000 years ago, everyone's skin was a shade of dark brown, protecting the body from the sun's intense UV radiation in sub-Saharan Africa. Various hues of brown evolved in response to the degree of UV radiation in particular regions of the African continent. Pushed along by natural selection, new pigmentation-changing mutations eventually established themselves in the local societies of genes outside Africa. Some of these mutations are complementary: populations in parts of Asia and Europe have skin that is equally light, yet the mutations responsible for that color are located in different genes. The corresponding alleles provide the same protection but result in slightly different shades, comparatively more pink or yellow.

It is difficult to think of a process necessary for building and maintaining your body that is not the result of a concerted effort of many genes. In many ways, the interactions among the genes are more crucial to our understanding than the genes are in themselves. Consider the thought problem of the Ship of Theseus, discussed by the philosophers of ancient Greece and applied to genetic interactions by the French geneticist Antoine Danchin. Every few years, one of the planks of Theseus's original wooden boat rots and needs to be replaced, until eventually none of the original planks remain (Figure 5.4).

Is this object, which had all its components replaced, still the same boat? Of course it is! What is special about the boat is not its individual planks; it is how these planks work together to form a vessel. Each plank is characterized not so much by the identity of its wood but by its position in the boat's design—in other words, by the planks to which it is connected.

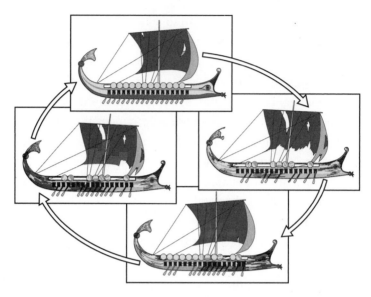

Figure 5.4: The ship of Theseus. If rotted parts of a boat are replaced until none of the original parts remain, is it still the same boat?

The relations among the objects, not the objects themselves, are what is important. Likewise, to appreciate the importance of each gene, we must look at its functional interactions with other genes. While you only have 20,000 genes, the number of interactions among them is much larger.

When genes collaborate to accomplish a function, whether skin color or a metabolic pathway, epistasis is at work—as we saw also in a set of genes that lead to the same disease when mutated. We also know that a single gene can cause more than one effect, a property called pleiotropy. Because of pleiotropy, a mutation in a single gene can affect several normally unrelated functions, causing a genetic syndrome: a group of traits or abnormalities that occur together and are associated with a particular disease.

As we saw, several genetic syndromes are Mendelian, showing that a single gene can exert influence on diverse processes. It is not unusual for a single mutation to cause a wide array of symptoms. An example is the genetic syndrome ataxia-telangiectasia, a disease caused by a mutation of the *ATM* gene. This syndrome affects the nervous system and the immune system and causes sterility, predisposition to cancer, dilated blood vessels, and extreme sensitivity to radiation.

Genes encoding enzymes, the proteins that catalyze chemical reactions, are often promiscuous, meaning that they are capable of breaking down different molecules—a special type of pleiotropy. One of many examples is the enzyme carboxylesterase 1, made by the gene *hCE1*. This enzyme is so promiscuous that it can break down a large variety of drugs, including cocaine, heroin, and methylphenidate (used to treat attention deficit disorder). A gene that has many functions is like an actor who appears in many movies: each role is different, but the actor is the same.

Somewhat ironically, an example of one gene having several seemingly unrelated functions was already present in the set of pea traits that Mendel chose to study. As Mendel wrote about one particular trait:

> To the difference in the color of the seed-coat. This is either white, with which character white flowers are constantly correlated; or it is gray, gray-brown, leather-brown, with or without violet spotting, in which case the color of the standards is violet, that of the wings purple, and the stem in the axils of the leaves is of a reddish tint. The gray seed-coats become dark brown in boiling water.

We now understand that seed-coat color must be determined by a pleiotropic gene, which also controls the color pattern of the flowers.

If a gene is promiscuous, then different mutations to it may affect each of its functions separately, compromising a body's integrity in seemingly unrelated ways. A case in point is the SOX9 gene. If a mutation destroys all functionality of the gene by fully abolishing its ability to produce proteins, the effect is a striking array of symptoms, including sex reversal, skeletal malformation, and a cleft palate. However, it is possible for only one or two of these symptoms to appear. Remember that genes possess a regulatory region, with molecular switches that control their activity (Figure 5.5). These switches are operated in rather specific ways. The SOX9 gene has three separate switches: one that turns it on in the testicles, another that turns it on in cartilage, and a third that turns it on in facial

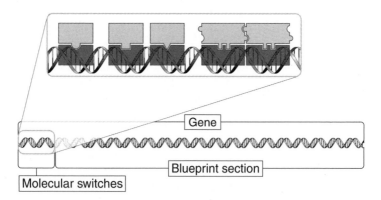

Figure 5.5: The top schematic represents the architecture of the regulatory region of the SOX9 gene. The dark gray shapes indicate sequences corresponding to molecular switches that are bound by specific proteins (light gray) and that regulate how much of the SOX9 protein is produced.

development. If a mutation damages one of these switches, then only the corresponding function of SOX9 is suppressed. As a consequence, mutations to one control region of SOX9 cause sex reversals, while mutations to another lead to a cleft palate.

Promiscuous Teams in Bacteria

Your genome built and controls your body, an organism consisting of several hundred different cell types interacting in myriad ways. We are still far from a comprehensive understanding of these interactions. Efforts are underway to analyze how the activity of genes is affected by promiscuity (pleiotropy) and teamwork (epistasis). A glimpse at what a genetic interaction map might look like has come from examining the much simpler genome that controls the bacterium E. coli, arguably the best-understood organism on the planet. E. coli's simplicity makes it relatively easy to work with and has led to many of the basic discoveries of molecular biology. Because the genome of E. coli has only about 4,000 genes, it has been possible to understand and characterize most of its parts and how they work together. Despite these discoveries, E. coli biology is a testament to how far biology is from any comprehensive understanding: the functions of about one-third of the genes of even this simple bacterium are still unknown.

One particularly well-studied subsystem of E. coli is its system of biochemical reactions—its metabolism. Almost every chemical reaction that occurs during the conversion of different nutrients into the building blocks of the next generation is catalyzed by an enzyme encoded in *E. coli's* genome. The map relating *E. coli's* genes to its biochemical capabilities shows well over 2,000 functions, established by more than 1,300 genes. Obtaining this map involved intricate biochemical detective work, often requiring years to reveal the function of just a single enzyme.

As in all organisms, metabolism in *E. coli* is a fairly promiscuous affair: almost half of the genes contribute to the catalysis of multiple chemical reactions. But epistasis is also at work: on average, every gene's protein product pairs up with those of two other genes to form a protein complex that then performs a specific function. It is likely that promiscuity and teamwork are even more prevalent in more complex societies of genes such as our own.

It is easy to see how multifunctional proteins can arise. Most enzymes have evolved to process one or more specific chemicals (their preferred substrates), but they also show some "accidental" activity with a range of other chemicals that may not normally be present in the cell. If the environment changes so that one of these non-preferred chemicals suddenly becomes available as a potential nutrient, then the already existing promiscuous enzyme provides an easy starting point for the evolution of new metabolic capacities.

Promiscuity and teamwork complicate the relationships between and among genes, but they are also at work in a single gene. The consequences of a mutation are frequently dependent on previous mutations in that same gene. An example of this complexity is *beta-lactamase*, a gene whose protein product is involved in *E. coli's* resistance to antibiotics. Until recent years, urinary tract infections were treated with the antibiotic penicillin. However, a substantial fraction of *E. coli* now carries five specific mutations in the gene for beta-lactamase that increase the normally only moderate resistance to penicillin by a whopping 100,000-fold. When "naïve" *E. coli* are subjected to penicillin, resistance evolves through the accumulation of the five mutations, one after the other. There are 120 possible orderings in which the five mutations might occur. However, some mutations even *decrease* antibiotic resistance if specific other mutations are not already present when they first arise. As a consequence, only ten of the 120 possible orderings lead to an increase in antibiotic resistance at each intermediate step (Figure 5.6), and these are practically the only paths that natural selection can follow. The individual mutations in the gene for beta-lactamase depend on one another; only as a team do they confer optimal antibiotic resistance.

A mutation to its gene can affect several properties of beta-lactamase. By studying the trade-offs among the different effects of a mutation (pleiotropy), we can understand how different mutations influence one another—that is, we can understand the origin of epistasis. The exchange of a specific letter in the second half of beta-lactamase enhances the protein's ability to destroy penicillin (good for the bacterium), but at the same time makes the beta-lactamase protein unstable, so that it easily falls apart (bad). Activity and stability are normally not related in beta-lactamase, but in this case, one single mutation affects both processes. In contrast, a second single mutation in the first half of the protein slightly reduces beta-lactamase's ability to destroy the antibiotic, yet boosts the stability of the protein. Neither of these two mutations is very helpful on its own, but when both mutations co-occur,

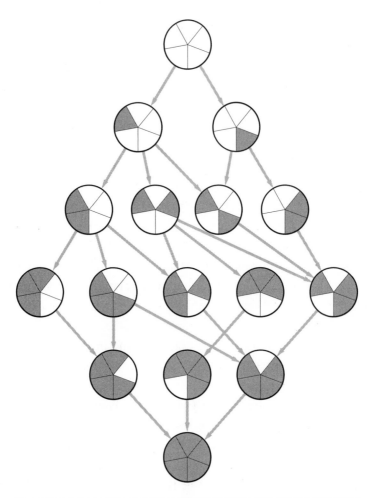

Figure 5.6: A map of the possible paths of accumulating the five mutations that together provide the bacterial beta-lactamase with a 100,000-fold stronger resistance to penicillin. Each circle indicates a beta-lactamase with a particular combination of the five mutations (dark gray). An arrow indicates the accumulation of one new mutation that increases antibiotic resistance. Because of epistasis, not all paths are possible: sometimes, a certain next mutation would reduce resistance, and only later mutations would increase resistance again. In such cases, no arrow is drawn, as natural selection would strongly disfavor individuals with such combinations of mutations.

they become a successful team: better at destroying penicillin because of the first mutation, and still stable because of
the second mutation.

The Perfect Drug

Doctors prescribe treatments on the basis of how effective they
have been. That is, they choose what worked better, on average, than other treatments in the past. This is an imperfect
science; the unique combination of alleles in your genome
means that the cause of your symptoms, as well as your reaction
to medications, can differ from those of other patients. There is
no one-size-fits-all approach to diagnosis and treatment.

In the 1950s, a naturally occurring molecule called succinylcholine was introduced as a muscle relaxant during major
surgery. It works well for the overwhelming majority of people,
but for some it proved life threatening. Normally, a protein
breaks down succinylcholine, and its muscle-relaxing effects
are washed out minutes after stopping the supply of the drug.
Unfortunately, some people lack a working version of the
corresponding gene. When taken off a breathing machine
quickly after the succinylcholine supply is withdrawn, their
chest muscles, necessary for breathing, may still be paralyzed.
Suffocation can result if the patient is not put back on the
breathing machine until the drug's effect wears off.

Hopefully, such risks may become a thing of the past.
The U.S. Food and Drug Administration, the government
agency responsible for the approval of new medicines, now ap-
proves drugs tailored to subgroups of people distinguished by

their genetic makeups. The goal of this approach, called personalized medicine, is to tailor drugs to individuals based on genomic evidence. This would make drug reactions more predictable, improve drug safety, and optimize treatment.

One example of an advance in personalized medicine is the drug tetrabenazine, a treatment for Huntington's disease, a condition caused by mutations of the *huntingtin* gene. The mutated *huntingtin* gene makes a defective protein that gradually damages cells in the brain, leading to a range of problems in muscle coordination, cognition, and mood. For tetrabenazine to be effective, it must be converted to an active form by the enzyme CYP2D6, but the amount of this enzyme varies from person to person. Doctors are now able to test how much CYP2D6 a patient produces and adjust the drug dosage accordingly.

Personalized medicine also holds promise for cancer management. Cancers that are traditionally grouped together may be caused by different sets of mutations, and each subtype thus requires its own treatment. Drugs can also be tailored to the genomes of viruses: recent treatments of hepatitis C are targeted to specific genomic types.

One day, we may have singled out the teams of genes involved in every important disease. Based on your genome, doctors might then calculate the probability that you'd be suffering from, say, a migraine by the age of thirty-eight. However, given the complex and promiscuous interactions in the society of genes, this scenario of an omniscient genomic medicine seems unlikely. Doctors will, for the foreseeable future, not be able to do away with more traditional methods. While this may not be optimal for our health and lifespan, it may benefit our psyche. Knowledge is not only power; it can also be a burden. If doctors told you that you had an 82 percent chance of suffering a devastating disease by a specific age but could not suggest an adequate cure, then this insight might represent more information than you need. The availability of genetic information leads to important ethical and philosophical issues fraught with dilemmas that will occupy us for some time to come.

The alleles of a society of genes are functionally entangled in a complex network. When comparing such a complex network with that of another society of genes, where would you expect the differences? Are such differences the cause or the consequence of the evolution of a new species?

6

The Chuman Show

You cannot step into the same river twice.
—Heraclitus

OLIVER DID NOT BECOME GREAT—HE HAD GREATNESS THRUST upon him. From his birth in 1976, Oliver entered stardom as the first "chuman"—half chimpanzee, half human (Figure 6.1). He preferred to walk upright, unlike his fellow chimpanzees, who loped along on all fours. He had no facial hair, which gave him a human-like appearance. In most respects, though, he was clearly a chimpanzee: no speech, no advanced use of tools, no evidence of complex thought. Still, for a few years, Oliver the chuman was a celebrity. With the rise of molecular tools, it was finally possible to confirm Oliver's nature. Was he truly the outcome of a sexual encounter between a human and a chimpanzee?

That question can be answered with some simple genetic accounting. Recall that human genomes are comprised of forty-six chromosomes: two twenty-three-chromosome sets, one inherited from each of our parents. The chromosome count of chimpanzees is slightly different: two twenty-four-chromosomes sets. Are chimps so different from us that they have an entire chromosome that we miss out on? No; our chromosome 2 turns out to be a combination of two

Figure 6.1: Oliver, who was on show in the 1970s as a human–chimp hybrid: a chuman.

smaller chromosomes in the chimpanzee genome, as shown in Figure 6.2.

Given that humans and chimps had a common ancestor about 6 million years ago, this difference in genome organization could have been due either to the breaking of a large, humanlike ancestral chromosome during the evolution of chimps or to a fusion of two smaller, chimplike ancestral

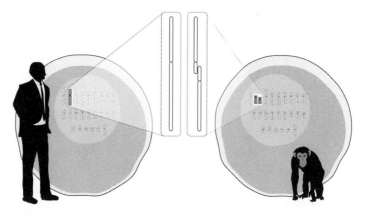

Figure 6.2: The genes on human chromosome 2 line up with two smaller chromosomes in chimpanzees. Our common ancestor had chimplike chromosomes, two of which accidentally got glued together as humans evolved.

chromosomes during human evolution. We now know that fusion, not breakage, is responsible for our different genome organizations. Every chromosome has one particular region called the centromere; when cells divide, molecular "ropes" are attached to this region to pull the two chromosomes of a matching set apart. Alongside its current centromere, the human chromosome 2 shows remnants of a second, former centromere, betraying the history of this chromosome as a combination of two ancestral chromosomes. Further support comes from gorillas and other more distant relatives of humans and chimpanzees, which also have the chimplike, nonfused versions of the chromosomes. Thus, the common ancestor of chimps and humans had a chromosome collection resembling that of the chimp and of other apes; at some point during human evolution, two of the chromosomes were fused together to form our current chromosome 2.

If Oliver were indeed the result of a mating between a human and a chimpanzee, he would have inherited one set of chromosomes from each of them. With twenty-three chromosomes from his human parent and twenty-four from his chimp parent, Oliver's genome should have been made up of an uneven number of chromosomes, forty-seven in total. Such a hybrid would present a big challenge not only for our legal system (would he have human rights?) but also for his genetic system. An odd number of chromosomes could not be rounded up in pairs, severely confusing the fair-coin-toss system of meiosis required to produce sperm. Consequently, Oliver's sperm production would break down, or else his sperm would be severely defective. For the same reason, infertility is indeed the fate of most hybrid animals, even in the rare cases in which hybrids are viable at all. Mules, for example, are the result of a mating between a male donkey (with thirty-one chromosomes) and a female horse (with thirty-two chromosomes), and are only rarely able to produce offspring themselves.

It turned out that Oliver did not have forty-seven chromosomes; he had forty-eight, like any chimpanzee. The chuman had been sheer fantasy. Oliver was an unusual chimp, yet a chimp just the same. But what really stands in the way of a chuman? Did chumans ever exist?

Genomes in Flux

There is good reason to believe that, even without the fusion that became human chromosome 2, a chuman could not exist.

At its most fundamental level, the hindrance to a chuman is a consequence of what a species really is: a society of genes. The genes and their different alleles in this society mix and mingle freely among themselves but only very rarely with those belonging to the societies of other species.

As we have seen, specific genomes are fleeting assemblies of alleles. Come back in, say, 121 years, and you will find an entirely different set of human genomes on this planet. While people and their specific genomes will perish, the alleles—the members of the society—will still be around. But over time, the genes change, too. New alleles of genes are born through mutations, and over evolutionary timescales sometimes rise to dominance over their ancestors. Entirely new genes join from time to time, while old ones that lost their contribution to the society's overall success in a changing world are kicked out. Even if the global society of genes—composed of all the genes and their alleles in one species—is much more stable than the set of alleles making up the specific genome of one individual, it cannot help but change over time. Societies evolve.

The society of genes might change because of environmental conditions, but it will evolve even if it doesn't need to adapt to a new situation. We've already seen how this happens: when your parents produced the sperm and egg cells that later merged to form your genome, a few new mutations arose, bringing new alleles into existence. Some of these alleles may have been identical to alleles already present in other humans; some may have been variants that appeared before and died out, and some may even have been completely new alleles.

Consider the fate of one of these novel alleles in your genome. Let's use as a hypothetical example an allele that carries the letter A at a particular position of the chromosome 5 you received from your father; at this position, all other human chromosomes 5 have a G. Your A-containing allele may not do so well in the coming generations, eventually falling out of existence again. As we saw for fly eye color in the Chapter 5, this outcome can come to pass by chance alone. If you had a single child, there would be a 50 percent chance that you would pass on the A on the chromosome inherited from your father and an equal chance that you would pass on the G from your mother. In the latter case, the A would be lost from the world. But in the former case, there would be a small chance that the A would eventually spread throughout the whole population. In many thousands of generations, it might be part of the genome of every single individual—its frequency in the society of genes would have reached 100 percent.

Because of the role of chance in sexual reproduction, no individual allele can remain at exactly the same frequency in the society of genes over time; its popularity in the society of genes must rise or fall each generation. That is what we call evolution when we consider the long-term perspective; it occurs at the level of the society of genes, not in any particular individual. The society of genes is the arena in which alleles compete.

The likelihood that the A allele would spread throughout the whole population is very slim, for the obvious reason that A is vastly outnumbered by the ubiquitous G alleles. The odds for A are related to the size of the population. In humans, the chances of A to rise to supremacy over the normal allele is one in 14 billion (A is outnumbered by the sum of two G alleles in every other human). You might think that, with such slim odds, no new allele is ever likely to spread through the entire population. But again, there is power in numbers. Every letter in the genome of a fertilized egg cell has a probability of one in 100 million to be newly mutated in the parental germ line. Given that the genome consists of about 6 billion letters, this means each and every genome contains about sixty new mutations. There are 7 billion genomes with sixty new mutations each, and every one of them has a one in 14 billion chance of spreading to supremacy. The billions cancel out, and thus, thirty new alleles replace their predecessors in the global society of genes in every generation, exactly the same number as there are new mutations in one individual half-genome.

This calculation shows that the society of genes must evolve at an appreciable rate, because countless new mutations arise each generation, and some of the resulting new alleles replace other alleles in the society. Mutation rates, too, are under genetic control. The society of genes maintains a balance between the dangers of too many mutations and too few. Take the case of the genes that encode our intricate repair machinery. An allele that caused too large a number of mutations would disturb the function of many genes needed to successfully build and manage the human body, and the carriers of this allele would not fare well. Too few mutations would be just as dangerous, because no adaptation can occur without genetic variation. As soon as the environment changed—and environments are also in constant flux—the carriers of alleles that suppress mutation rates too much are in trouble. Mutations are troublesome but necessary. No pain for the individual, no gain for the society.

The letter sequences of the genes of humans and chimps are almost 99 percent identical. In addition to the 1 percent difference in the genome sequences that we both inherited from our common ancestors, humans and chimps also differ by almost 3 percent of DNA that is found exclusively in humans or in chimps; this is akin to the difference between individual humans, which is 0.1 percent when focusing on single letter changes but is 0.5 percent when also considering insertions and deletions of DNA.

Assume then that the G allele is common to both humans and chimps. Its eventual replacement in humans with your newly arisen A allele will constitute another difference between the two species. Each of the differences that amount to the 4 percent that now distinguish the genomes of chimps and humans arose in this way. Each change first appeared as a mutation, either in one of our human ancestors or in an ancestral chimp. Most differences between humans and chimps likely occurred by chance. In some cases, though, when a new allele provided an advantage, its spread through the population was speeded up by natural selection. One mutation at a time, the two species became more and more different. They diverged, little by little, slowly but surely.

The Key That Jams the Lock

Two individuals with genomes that are 99.5 percent the same—for instance, those of two humans—can have children, yet a human and a chimp—whose genomes are 96 percent identical—apparently cannot. Where is the breaking point? How much difference is too much?

Populations meeting up for the first time after a long separation can often successfully breed with one another. You can quite accurately predict whether they can or cannot from the amount of time that has passed since they split up. If two local populations of animals are rejoined after having been separated from each other for 1,000 years—say, by the emergence of a river that eventually dried up again—they are more likely to produce fertile progeny than if they had been separated for 10 million years. The more time passes, the more different the two societies of genes will become, and mixing them again will become more difficult.

The making of fertile offspring is not an all-or-nothing affair. A new encounter of previously isolated populations may, for example, result in a 50 percent chance of offspring dying before reaching maturity. As the time these populations are left in isolation increases, the risk increases, until they can no longer produce any viable offspring at all. The two isolated groups become more than just distinct and isolated populations: they are different species.

Darwin titled his revolutionary book *On the Origin of Species,* but he didn't have the information that would have allowed him to understand how new species actually form. Today, we know that the society of genes is at the heart of this process. Most of the genomic changes occur through the random process we encountered as "drift" in Chapter 4, with an important caveat: each change to a gene could only be one that did not hurt its carrier. If it did, the corresponding mutation would quickly have been lost from the society. In other words, each new mutation that is not immediately lost needs to be compatible with the existing alleles of other genes. But as soon as that mutation becomes ubiquitous, the following mutations need to be compatible with the reality of the new society, which now includes the previous mutation. Thus, accumulation of mutations has a historical aspect. The spreading of one specific mutation may promote, or may exclude, the rise of following mutations. To understand why a chuman is impossible, it is important to consider that evolving populations accumulate a series of changes that are compatible with one another but not necessarily with the gene versions present in their ancestors, let alone with changes that simultaneously arose in other populations.

This is how things fall apart: imagine that the two populations that had been separated by the river are reintroduced to one another after each one has accumulated 1,000 mutations that replaced older alleles in their society of genes. In a species like ours, it might take on the order of 100,000 years for this accumulation to take place. At that point, the two societies would differ in 2,000 places: about 1,000 due to mutations on one side, and another 1,000 stemming from mutations on the other side of the river. If a member from one population impregnated a member of the other, their baby would inherit all 2,000 mutations, combined into one genome. This will be the first time that the 1,000 changes of one population find themselves in the company of the other population's 1,000 mutations. These sets of mutations have never been tested against each other. They did not accumulate slowly in succession; instead, they are combined all at once. Will they gel? The chances are slim that all of these mutations are fully

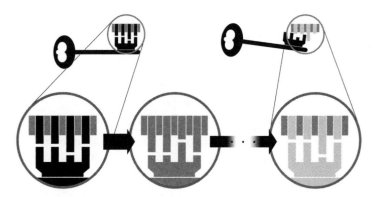

Figure 6.3: A succession of lock / key combinations, showing how matching changes in lock and key accumulate over time: that is, they co-evolve. After a few such steps, the original key (black) will no longer fit the evolved lock (light gray).

compatible. Had it been 10,000 different mutations in each population, the chances would have been slimmer still.

Here's another way to think about this. Imagine a lock and its key, where the shape of the lock and that of the key can change over time, and the usefulness (the "fitness") of the lock / key system is determined by how well it operates (Figure 6.3). The key can modify its existing teeth; in an analogous fashion, the lock can change its parts that interact with the key. With time, a random change may occur to a part of the key, making one of the teeth longer. The lock will still work (otherwise the key would not have been allowed to change), but its operation may not be quite as smooth as before. Some time later, a change might occur to the lock that better fits the new key shape; this change now would be beneficial, and thus more likely to persist. As more and more such changes accumulate, the lock and its matching key will become increasingly different from their original shapes. If you would take the original key and tried to open the evolved lock, the key would likely jam the lock.

The lock and key in this analogy correspond to different genes that interact with one another in one society of genes. Remember that such interaction is ubiquitous; for example, many proteins need to attach themselves to other proteins to perform their function. Some of the mutations that arise during evolution change the shape of proteins. A small change to one protein (the key) might slightly reduce its ability to attach to its partner (the lock). This minor change would, in turn, increase the probability that a random mutation causing a corresponding change in the partner would also establish itself in the population. When this process continues for a long time, the interacting parts accumulate coordinated changes: they co-evolve.

Evolution is contingent on the molecular history of the society of genes, and if nature's evolutionary experiments were repeated, you would be almost guaranteed an outcome that differs in its details—each evolutionary change is, after all, also an outcome of chance. Thus, when two populations evolve in isolation and then come back together for the first time, a mess ensues: the members of the different societies of genes do not know how to interact anymore.

A Touching Family Reunion

Oliver was no chuman, and it is highly unlikely that any chuman existed in recent times. But there is good evidence that chumans once lived, at least in a sense. The 4 percent difference between humans and chimps should be equally distributed across the chromosomes: all chromosomes accumulate mutations at an equal rate, with the Y being the only exception, due to its sexless existence. However, that is not what we find when we take a closer look: the human and chimp X chromosomes harbor about 20 percent fewer differences than the other chromosomes. This reduced number of changes is unique to a comparison with chimps. If you compared your genome against that of the gorilla instead, all chromosomes, including the X, would show a similar amount of changes.

What does this teach us about how humans and chimps might have become separate species? About 6 million years ago, our ancestors and those of the chimps belonged to the same species. Then one day, a group of members of this species split off and set up a new home in an isolated place. They didn't look back. From that day on, the two lineages evolved in isolation, eventually becoming different species. In this scenario, all parts of our genome should show an equal degree of similarity to the chimp genome, since these all had the same amount of time to accumulate differences.

So to explain different degrees of similarity for different parts of the genome, we need to accept that there must have been sex between the two lineages long after they had split up, when they had already accumulated a sizeable fraction of the genomic differences we see today. At that time, we can think of them as early humans and early chimps, though it is likely that they both had the same number of chromosomes as their common ancestors—two times twenty-four.

The scandalous interactions between individuals from the two camps resulted in the infusion of chimp genes into the human lineage and/or vice versa.

How does this explain that we see more similarity on the X chromosome than on the other chromosomes? The simplest explanation may be an overall asymmetry in the matings: if equal numbers of males and females from the chimp lineage infused their genomes into the human lineage, then the X should harbor just as many hybrid regions as the other chromosomes. But if only the women were allowed into the early human villages, things would be different: all the matings that had an impact on the human society of genes would have been between a chimp-lineage woman and a human-lineage man. The daughters resulting from these couplings would have genomes that were exactly half human and half chimp. The sons would all have a human Y from their father and a chimp X from their mother. The human Y chromosomes would show no signs of the chimp mothers whatsoever, but because two-thirds of the children's X chromosomes from mixed marriages would stem from the chimp mother, the human X chromosome would show stronger remnants of this mixing than the other chromosomes, even many generations on.

These scandals happened a long time ago. As we explained in the context of Oliver's alleged half-human ancestry, your genome now is so different from that of a chimp that there is no longer a gray area: chimps and humans are without a doubt separate species. Are there individuals who are evolutionarily closer to humans than the chimps? No such creatures exist: today's humans are an isolated species. But things were different in the not-so-distant past. Up until only 40,000 years ago, Neanderthals were our next-door neighbors in Europe and the Middle East. In fact, Neanderthal and human bones from the same period were found in the Kebara cave in modern-day Israel, suggesting coexistence.

Neanderthals and the human lineage split more than 300,000 years ago in Africa. Soon after, the ancestors of Neanderthals emigrated to the Middle East and Europe, where *modern* humans encountered them again when they reached Europe much later. Thus, the genomes of humans and Neanderthals had ample time to diverge from each other in isolation: when modern humans arrived, they looked (and probably felt) quite distinct from their stockier, cold-adapted cousins. Were these odd-looking people human? Though we cannot be sure that it happened among consenting adults, there certainly were sexual encounters between members of the two groups (Figure 6.4).

How do we know about these encounters? While the exact details remain in the shadows, the outline of that story is told in our genomes. With great care, DNA can be extracted from ancient bones, giving us access to the Neanderthal genome. Because, as bears repeating, the society of genes cannot sit still, your genome is quite different overall from that of a Neanderthal. Interestingly, though, if you happen to be of recent African ancestry, then your genome is even a bit more different from that of a Neanderthal than those of non-Africans (that is, those individuals whose ancestors left Africa in prehistoric times). This systematic difference between Africans and non-Africans could not be explained if there were no further

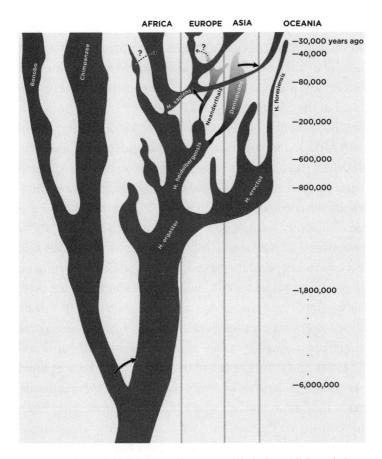

Figure 6.4: The evolutionary tree of humans and their closest living relatives, chimps and bonobos, with a detailed depiction of recent evolutionary relationships among groups of human ancestors. The numbers indicate how far in the past the events on this tree took place. The lineages leading to humans and chimpanzees split about 6 million years ago, although there is evidence that isolated sexual encounters between members of these groups occurred much later (arrow near the root of the tree). About 300,000 years ago, modern humans split from the ancestors of Neanderthals and Denisovans. The modern humans initially stayed on in Africa, and their cousins crossed into Europe and Asia. It was there that some modern humans reencountered them after they left Africa less than 100,000 years ago. The arrows connecting Neanderthal and Denisovans to human lineages indicate sexual encounters still recognizable in human genomes from the corresponding geographic areas. Redrawn with modifications from Lalueza-Fox and Gilbert 2011.

encounters between modern humans and Neanderthals after we split into two separate lineages more than 300,000 years ago. The difference shows that after the first waves of humans left Africa, they mated with the Neanderthals they encountered in the Middle East and Europe.

The presence of Neanderthal DNA in the genomes of non-African humans clearly indicates that much less than 100,000 years ago, humans were perfectly capable of producing successful offspring with Neanderthals. Despite our technological advances, it is generally believed that as a biological species humanity has not changed much since then. That our ancestors and Neanderthals mated successfully means that they were not separate species. Neanderthals were human: they formed a tribe that had lived in isolation for a long time, although not long enough to accumulate mutations that made their society of genes incompatible with ours. For the sake of simplicity, we will continue to use the terms "humans" and "Neanderthals" for the rest of this chapter, but keep in mind that the Neanderthals were part of the human species: distant cousins of those humans who left Africa in prehistoric times to venture into Europe and Asia.

Neanderthals were not the only human tribes already living outside Africa when our modern human ancestors arrived. Parts of Northern Asia were inhabited by distant relatives of Neanderthals, humans we call "Denisovans" after the cave in which their bones were first found. Mirroring the sexual contact between Neanderthals and early modern Europeans, Denisovans had sex with the humans who reached Southeast Asia, as reflected in today's Southeast Asian genomes.

Better Than Sex

So unless you are of recent African descent, your genome contains alleles obtained from ancient human races. But your genome is devoid of genes recently obtained from other species like modern chimps, gorillas, or orangutans: the impossibility of such mixing is what makes them different species. Defining a species is rooted in the role of sex: if your genome can mix with another one without noticeable problems for the offspring, then the two of you belong to the same species—your genes belong to the same society of genes. What about species that make do without ever having sex? Bacteria vastly outnumber more complex life forms, such as animals, plants, and fungi, yet they live woefully sexless lives. How can we know whether two bacteria belong to the same or to different species?

Until recently, we defined bacterial species according to vague and somewhat arbitrary concepts based on similarity of appearance or genome. These classifications were often rather generous. As an example, two bacteria classified as belonging to the species *E. coli* can be vastly more different in terms of genome content than, say, you and a dolphin. However, it turns out that even in the absence of sex, the definition of a species as a group whose members can mix with one another successfully also applies for bacteria.

In its essence, sex is about mixing genomes, allowing the alleles of the society of genes to form new coalitions in each generation. Bacteria form these new coalitions chastely (Figure 6.5). A bacterium takes pieces of DNA that encompass at most a few dozen genes from another bacterium and introduces these genes into its own genome. The new DNA can

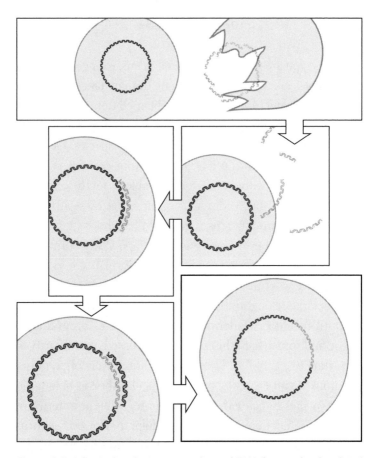

Figure 6.5: A bacterium integrates a piece of DNA from a closely related bacterium. The bacterium takes up a fragment of DNA from a distant cousin that died in its vicinity. The fragment aligns itself to the corresponding section of the bacterium's genome and replaces it in a process analogous to the recombination that exchanges matching chromosome sections in the production of human sperm or egg cells. This mixing of DNA in "bacterial sex" is less tightly regulated than in animal sex, but it provides the same service to the society of genes.

get into the bacterial cell in several ways. Once inside, the foreign DNA can be efficiently integrated into the bacterial genome by a recombination process similar to that used by our own sperm- or egg-producing cells to mix the chromosomes anew each generation. In animal sex, homologous recombination requires that the cell's machinery identify corresponding regions in the two matching chromosomes inherited from mother and father. In the bacterial equivalent, the same requirement holds: to be integrated, the foreign DNA must be flanked by stretches of letters that almost perfectly match corresponding stretches in the bacterium's own chromosome. This gives the species definition in bacteria an exceedingly simple interpretation: when the matching parts of two bacterial genomes are more than roughly 99.5 percent identical, they can successfully recombine. The genes of these bacteria belong to the same society of genes, and thus the bacteria belong to the same species. Because of the central role that homologous recombination plays in sex, it may be no accident that from humans to bacteria, the genetic difference that characterizes a species is of the same order of magnitude.

Make Love, Not War

From looking at our genomes, we know that when modern humans and Neanderthals met, they were perfectly able to mate. Neanderthals, after all, were human. Why then are Neanderthals no longer around? The Neanderthal groups may simply have merged with those of their newly arrived cousins from Africa. However, based on what we still see in much of the world around us, another scenario appears to be more likely. From the point of view of the Neanderthals, the human out-of-Africa journey was an invasion. For their part, the modern humans most likely viewed the Neanderthals as a threat or at least an annoying competitor for food and housing. In many encounters, our ancestors may have tried to kill Neanderthals. And if they did, it seems they were quite efficient, since the youngest Neanderthal bones we find are at least 40,000 years old.

This scenario describes just another round of the endless cycle of us-against-them racism, driven by the same types of genes and ideas that are still at their destructive work around the world today. Our distant ancestors crossed the Sahara to conquer the world, pretty much like the European sea powers in more recent centuries conquered large parts of Africa, Asia, Australia, and the Americas. The first modern humans in Europe and Asia likely raged war against the local people, now called Neanderthals and Denisovans.

But, similar to more recent invasions, some mixed couples of modern and Neanderthal humans made love, not war, firmly inscribing the Neanderthal genomic heritage into the modern human society of genes. The same seems to have happened throughout the world. There is, for example, also genomic evidence for sex between Denisovans and Neanderthals.

What were the consequences of the matings between modern humans and Neanderthals throughout Europe? For most parts of the genome, it probably did not matter much if a stretch of modern-human DNA was replaced by its Neanderthal counterpart. However, Neanderthals, having already lived in Europe for 200,000 years before modern humans arrived, were much better adapted to the local climate and to its pathogens. Their societies of genes contained specific alleles that significantly boosted the odds for humans to survive in these regions. Modern humans who inherited these adapted alleles from Neanderthal parents or grandparents had a competitive edge over their non-interbred neighbors.

HLA-A and HLA-C are two genes important for the functioning of the immune system. The proteins they encode are responsible for bringing protein fragments from the cell's interior to the cell surface, where they present them as the "All is well" or "Help, I've been invaded!" signals discussed in Chapter 2. The human society of genes contains many alleles of these two genes. The exact sequence of your copies influences which protein fragments will be brought to the surface of your cells—in other words, which pathogens you can recognize. If you are of Eurasian ancestry, chances are high that your genome contains versions of HLA-A and HLA-C that your ancestors obtained via sexual relationships with Neanderthals.

In a very similar vein, the corresponding HLA genes of many modern Asians are highly similar to those found in the Denisovan genome. Overall, the genomes of modern Asians have a 70–80 percent chance of containing an HLA-A allele that originated in one of the two ancient non-African populations—the Neanderthals and the Denisovans. In modern Europeans, the chance of having a Neanderthal allele is about 50 percent. Conversely, if your genome happens to be of African descent, your chance of having one of these ancient HLA-A types is only 6 percent, and these rare cases are probably a consequence of migrations back from Eurasia into Africa. While modern Europeans are not half Neander-thal, an important part of their immune system thankfully is.

The society of genes is always evolving. When a society splits into two, the societies cannot help but diverge irrevocably. To evolve a new ability, such as a larger brain, a species does not necessarily need additional genes; change is much more often driven by managing the same genes in a different way.

7

It's in the Way That You Use It

To do great things is difficult; but to command
great things is more difficult.

—Friedrich Nietzsche

HAVE YOU HEARD OF THE MARSHMALLOW CHALLENGE? HERE'S
how it works. The participants split into groups of three or four
people. To get a marshmallow as high up above the ground as
possible, each group gets twenty sticks of spaghetti, a yard of
tape, and a yard of string, and has exactly eighteen minutes to
complete the challenge (Figure 7.1).

Most groups never manage to get the marshmallow off the
ground. MBA students are notoriously bad at it. After spending
most of the time jockeying for power within their groups to
decide who will be the leader, they enter crisis mode seconds
before the end, when the marshmallow is finally mounted
to a flimsy structure. CEOs are not much better, but when a
project manager is added to a group, the likelihood of success
increases dramatically. Interestingly, kindergarteners do very
well at the challenge. Their secret? They start with the marshmallow and then look for ways to add to a structure that gets
it incrementally higher and higher.

The marshmallow challenge exposes the crucial role that
the art of management plays when accomplishing a task. Given

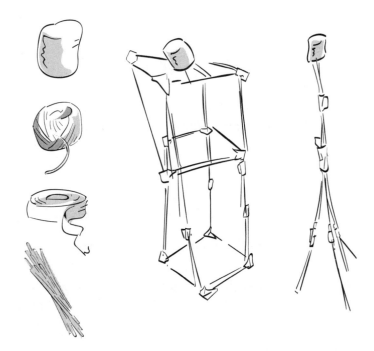

Figure 7.1: The marshmallow challenge: Take one yard of string, one yard of tape, and twenty sticks of spaghetti and build a structure that supports a marshmallow as high above the ground as possible. An huge variety of structures can be built from a small number of parts.

the same amount of tape, string, spaghetti, and time, many different outcomes are possible.

Speak Up

We humans have a number of biological innovations that set us apart from other animals, including from our closest relatives, the chimpanzees. We walk upright, have large brains, and invent technology. The most powerful human innovation of all may be speech. Our ability for complex communication by means of subtle modulations of air pressure may well be at the root of our ability for complex thinking, making speech fundamental to our understanding of the world, as the philosopher Ludwig Wittgenstein suggested. Where does speech come from? Which innovation in the society of genes was necessary to enable us to talk?

To answer this question, we need to identify the genes involved in creating and interpreting speech. One way to identify genes involved in language formation is to compare the genomes of people with a specific speech deficiency with those of people without that deficiency and look for consistent genetic differences between the two groups—the same GWAS strategy we discussed in Chapter 5.

A large family in England proved ideal for such a study. The grandmother had a severe language disorder that affected her ability to use and understand grammar and form comprehensible speech. Four of her five children were similarly affected, and of the children of these four, roughly half had the disorder. In the course of teasing apart the genetic differences between the genomes of the affected family members from those unaffected, a culprit emerged: a gene by the name of *FOXP2*. Those family members with the disorder had numerous mutations in and around the *FOXP2* gene.

Two additional reasons make *FOXP2* likely to be a gene important for speech. The first reason is the type of job *FOXP2* has: it is a manager, rather than an operator. You will recall that in the society of genes, certain members (the operators) carry out the work needed to run a cell—such as unwinding

DNA, building cell membranes, breaking down sugars—while others (the managers) regulate the operators. Most managers belong to a family of genes called transcription factors: these turn the expression of other genes in the cell on or off by binding to the genes' molecular switches (see Figure 5.5). The number of such transcription factor managers is determined by the size of the society they regulate: a genome with twice as many genes needs four times as many regulators. Humans have a large number of genes, and about one-tenth of these are transcription factors, including FOXP2. Speech is a complex trait, involving important changes in the brain as well as in the throat, so FOXP2 needs a large number of operators to carry out these tasks. However, because it is a general manager, if it stops doing its job, the operators also stop working, and speech is disrupted.

FOXP2 also seemed to be important for human speech because of its genomic neighborhood. Recall that when we compare two human genomes, we expect about one difference in 1,000 letters (ignoring missing and inserted letters). However, the region containing FOXP2 is exceptionally similar across humans. There are even fewer differences than the usual 0.5 percent, a uniformity that would be highly unlikely to occur by chance. Such a low number of differences would be expected only if nearly all letters in a region were vital to the fitness of the organism, meaning that any mutation in the region would be fatal to the carrier. As we saw, though, most changes in and around a gene are of no important consequence for its function. So why then would a piece of the genome be so uniform across all of humanity?

The pattern is indicative of something called a selective sweep. Up to one point in prehistory, our ancestors' communication must have been like that of other mammals: a small collection of simple calls. But one early human—let's call him Orpheus—must have had a mutation that enabled him to express more complex phrases. Those of his children who inherited the mutated allele would have been able to communicate with an unprecedented complexity when they spoke to their father and to one another. Because better communication allows improved cooperation, Orpheus's children and grandchildren would have been more successful and would have had more offspring than others in their community. In this way, by natural selection, Orpheus's mutation would have spread through the prehuman society of genes. After a few dozen generations, almost everybody within marrying distance would have inherited Orpheus's allele for improved communication.

Orpheus's children and grandchildren would not all have had the favorable mutation, and those who did would not have inherited Orpheus's entire genome; half their genome would have come from their mother. What would increase in popularity would not be Orpheus's full set of genes, but only the mutation itself. However, a mutation does not exist in isolation. It is a single letter that has a fixed place on the genome, it has neighboring letters and neighboring genes, and as the mutation gains in popularity, so do all of its immediate neighbors. This is because those very close neighbors are unlikely to be quickly separated from the mutation by the few recombination events that occur in the preparation for each new generation.

In other words, when a favorable mutation rises in popularity due to natural selection, its immediate neighbors on the genome typically come along for the ride. Soon, each member of a population will share the mutation and its neighboring alleles; eventually, that entire region of the genome will be identical in everyone. The trademark of a recent change due to natural selection is this lack of variation at the favorable mutation site itself and in the surrounding region. The stretch of your genome that contains *FOXP2* appears to be a result of precisely such a selective sweep.

The *FOXP2* gene confers the ability of speech on humanity but not on other mammals or birds, all of which have a version of the gene. *FOXP2* is a promiscuous gene; it plays many roles in the embryonic development of organs common to all mammals and birds. So what kind of change to *FOXP2* bestowed Orpheus and his descendants with superior communication skills?

The answer lies not in what *FOXP2* is, but in how it is used. While *FOXP2* is a manager, it also has to be managed. *FOXP2* is turned on by other managers at a set of predefined times and locations in the human body. For example, it is activated during the development of the lungs and the gut. In contrast to the version possessed by chimpanzees and other apes, human *FOXP2* is also expressed in "area X," a specific region in the brain—the one that neurologists hold responsible for speech. It appears that our society of genes did not need a new member to enable speech. The invention of speech arose from a change in management rather than from the acquisition of new tools.

Returning to Orpheus, we now have some insight into the mutation responsible for his speech. The mutation to the DNA sequence of *FOXP2* did not change its function but rather it changed the way that other proteins bind to it. This in turn changed the regulation of when and where *FOXP2* would function. This change is analogous to the marshmallow challenge, where a highly successful design results from a better strategy for working with those same parts available to all groups.

Birds cannot speak, but to some extent a birdsong is to birds what speech is to us. Birdsongs are longer and more complex than simple calls, and they are associated with courtship and mating. They have their own grammar, a structure that often resembles that of human music in its diversity of expressions and its rhythmic regularity. Songbirds in many species learn at least part of their songs from their fathers, leading to the development of local dialects, analogous to human speech.

Not all species of birds sing. What then are the differences between those that do and those that don't? There is no "birdsong gene" that is missing from the genomes of nonsinging birds. But singing birds have *FOXP2* expressed in area X, the part of their brains that corresponds to your area X and that has been mapped for its importance in song learning in birds. Moreover, canaries change their songs in certain seasons, and it is only then that *FOXP2* is activated. The similarity between birdsong and human speech is striking: the same change in the management of *FOXP2* is crucial for both species. The action of *FOXP2* alone cannot explain the complex trait of language, of course, but the evidence is compelling that expression of this gene in a particular brain region is a prerequisite for language and complex grammar.

Big Brain Theory

Thanks to our large brains, humans have been able to develop and master ever more complex technologies, from harnessing fire to building smartphones. Does building a larger brain require the introduction of new genes into the society of genes? The FOXP2 story suggests that changes in management may be enough. One way to make a larger brain would be to let the brain cells divide for slightly longer as the brain develops, generating more cells in the process. Our brain is in many ways like that of chimpanzees; what may have evolved in the 6 million years since our gene societies went their separate ways is how the genome's managers regulate brain development.

There is some good evidence for this idea. One difference between the genomes of chimpanzees and humans occurs in the region that contains the molecular switches for a gene called GADD45G, which is a manager-type gene. GADD45G has been labeled as a tumor suppressor gene, because it is involved in managing which cells should stop growing, a crucial task in the suppression of cancerous tumors. In the human version of GADD45G, a sizeable chunk of sequence—3,200 letters—is missing in the DNA that makes up the regulatory region of the gene. If the corresponding genomic region in mice is removed, one of the genes involved in brain growth changes its expression. Thus, one plausible scheme for how we came to have larger brains is that one manager gene lost its ability to tell a specific brain region to stop growing at a certain time during embryonic development. Interestingly, Neanderthals, who also had large brains, shared this deletion with modern humans.

In explaining differences between humans and chimpanzees, changes in management appear to be the rule rather than the exception. There are practically no genes that are unique to us or to them. Moreover, the few differences in our genes' amino acid sequences are mostly of minor consequence to the protein's function. We both have virtually the same operators and managers, but the managers give different instructions.

This notion recalls an urban legend about the toothpaste company Colgate. Many years ago, faced with dwindling sales, the company hotshots held a meeting to brainstorm for ideas that might boost sales. A cleaning lady, who happened to be in the room, suggested making the tube's hole bigger, so that more toothpaste would come out at each use. The rest is history. It wasn't necessary to make a different product; just a tiny change made a big impact.

The genome of each of your cells encompasses 20,000 genes, resulting in a mind-blowing range of different genomic activities. Put simply, a cell can turn the state of each of its genes on or off: either the gene is read and its protein is produced, or the gene remains unread and lies dormant. The number of different possible states of genome activity is, for all practical purposes, infinite, even if not all of those states are viable. Consider electric circuits, in which the same resistors and capacitors are wired together in one way to form a fire alarm and another way to form a radio (Figure 7.2).

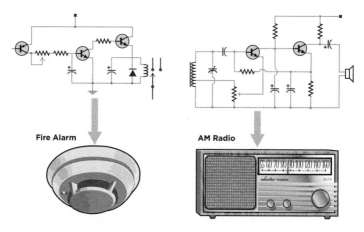

Figure 7.2: The same elements of a circuit can be wired differently to achieve radically different functions.

The function of the various cells in your body follows from their pattern of gene activity. While each of the cells has virtually the same set of genes, not all of those genes are "on" at any given time. For example, a specific type of liver cells will have its own on / off configuration: only those genes required for producing this liver cell type will be active, while all other genes will be switched off. Through changes in the pattern of management—the exact configuration of which genes are on and which off—your genome encodes the many different types of cells in the body and would theoretically be able to control a myriad more that differ from those already present. There is hence often no need for the invention of novel genes.

A more accurate comparison between humans and chimps would be one based not only on a census of their genes but also on a census of on / off switches across different cell types. When you carry out such a comparison for brain, liver, and blood cells, you find that the differences in gene-expression management between humans and chimp are most striking in the brain. This should come as no surprise, since the brain is the organ that most distinguishes us from other animals. Our expanded intellectual capacities may then indeed be a consequence of changes in gene management.

Genetic Turn-Ons

How exactly do the managers let other genes know what they're supposed to do? In science, the key to discovery is often reductionism. As biologist Peter Medawar famously said: "Science is the art of the soluble." A complicated process is made up of many mysteries, too many to be solved all at once. The trick is to focus on only one mystery at a time.

Accordingly, when looking for an answer to the question of how genes are managed in the dizzyingly complex human, we have to reduce the problem to manageable bits. To start making progress, let's examine one single on / off switch in our old friend *E. coli*, a much simpler organism than we are.

Consider *E. coli's* switch involved in digestion of lactose, the main sugar in milk. *E. coli's lac* operon is a genomic region that contains a set of genes that together encode the necessary machinery to take up and digest lactose. The activity of this set of genes needs to be under tight regulation because individual *E. coli* cells face severe competition from surrounding bacteria. The genes need to be active when needed. On the other hand, bacteria cannot afford to waste energy or the limited space of their workshop, the cellular interior, on proteins that are superfluous. Consequently, the *E. coli* genome has evolved to regulate its operons according to the available resources.

The *lac* operon contains a region of on / off switches, which regulate the expression of its whole set of genes (which are managed and read together). The management of the *lac* operon responds to the presence or absence of lactose or glucose. If lactose (but not glucose) is available in the environment, the *lac* genes have to be activated so that they can help convert the lactose into cellular energy; conversely, to avoid wasting resources, the genes have to be deactivated when there is no lactose around. When glucose, a more nutritious food source, is available, it pays for the cell to put its whole budget into producing glucose-digesting proteins. In that situation, it has to shut down the production of proteins useful for lactose digestion, even if lactose is present. In short, the lactose-digestion machinery—made from the *lac* gene set—should be activated only when lactose is available and nothing tastier is on offer.

How is this program encoded genetically? As you'll recall from Chapter 1, the polymerase is the protein machine that reads a gene's sequence into messenger RNA, thereby effectively turning that gene on. To read the *lac* genes, the polymerase first needs to attach itself to the beginning of the *lac* operon. When no lactose is present in the cell, a repressor protein (a low-level manager) sticks to the DNA in front of the *lac* operon, right where the reading machinery is supposed to jump onto the chromosome. Since the repressor is in the way of the polymerase, the *lac* genes cannot be read, and the machinery for processing lactose is not made. When lactose molecules reappear in the environment, some of them drift into the cell and attach themselves to the repressor proteins. Their

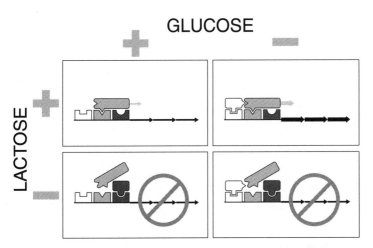

Figure 7.3: The logic gate encoded by the *lac* operon. Left and right columns show situations in which glucose is present and absent in the bacterium's environment, respectively; the top and bottom rows represent situations in which the alternative sugar, lactose, is present and absent, respectively. The three *lac* genes (black arrows) are needed to digest lactose. When lactose is absent (bottom row), a repressor protein (dark gray) prevents expression of the *lac* proteins, which are of no use without lactose. When glucose, the preferred sugar, is absent (right column), an activator (white) encourages the binding of the polymerase to boost gene expression of the *lac* genes. The *lac* genes are only highly expressed when glucose is absent and lactose is present (top right).

adherence slightly alters the shape of the repressors, rendering them incapable of clinging to the DNA. With release of the repressor proteins, the polymerase gains free access to the DNA, and the proteins for processing lactose are produced (Figure 7.3). This is the first part of the management algorithm: no lactose, no turning on of the *lac* genes.

The polymerase may find its starting point for reading the *lac* operon just by chance, but this occurs at a very low rate, too slow to produce appreciable amounts of the proteins needed to digest lactose. To direct the polymerase to its intended place of action, a second manager produces an activator protein, which binds to a docking point just in front of where the polymerase needs to attach. However, when glucose is available in the cell, the activator protein itself becomes inactive, and only very low amounts of the lactose-digesting machinery are produced. This is the other half of the algorithm that determines *lac* operon expression: to take full advantage of the more nutritious glucose, the cellular resources need to be diverted from the processing of lactose. The management of the *lac* operon thus works like a simple logic gate in a computer: the *lac* proteins are produced if lactose is available and glucose is not; in all other situations, these proteins are not produced in appreciable amounts.

The central processing unit, the "brain" of a computer, is built up from many millions of such simple logic gates. The same type of calculations as those performed by each logic gate can be executed by the genome, using the principles we saw for the *lac* operon. The transcription managers transmit signals from their surroundings to specific places in the genome. Logic gates are constructed through combinations of transcription factors that either induce or hinder access of the transcription machinery to the managed gene.

Successful management in the society of genes is not based on intelligence or intentions. The dance of the management proteins along the chromosome is solely a consequence of their molecular affinities: due to their shape and the electrical charges on their surface, the proteins attract specific molecules or are themselves attracted to larger molecules, such as specific sequences of DNA letters. Looking at the *E. coli lac* operon allows us to gain some fundamental understanding of how management is executed in our own society of genes, which encompasses many more layers of complexity.

In Chapter 5, we encountered the gene SOX9. When its protein-making capacity is mutated, many phenotypes result. Many managers bind SOX9, leading to its induction or repression. In the case of *E. coli*'s lactose genes, a complete set of genes is regulated together; in your genome, each individual gene gets its own computing units. Complex networks of management can be built up through this process. Whole sets of genes that function together are managed by cascades of transcription factors. In this way, the activity of a transcription factor itself is managed by transcription factors, enabling complex information-processing chains.

To get a glimpse of a small section of the network, let's take another look at the SOX9 gene and focus on the determination of your sex when you were an embryo. Initially, SOX9 was expressed regardless of your sex. If you are a woman, your genome started to also express a protein called "beta-catenin" in the part of your developing body that was to become your ovaries. Beta-catenin proteins found and stuck to SOX9 proteins, in a suicide mission that marked them both for destruction. With dwindling SOX9 levels, these cells started to build ovaries. To make sure that this developmental decision cannot be reversed, other management proteins keep SOX9 levels down in your ovaries by blocking the transcription of SOX9 for the rest of your life.

If you are male, your genome contains a Y chromosome, and things took another turn. Your Y chromosome harbors the *Sf1* gene, which further boosts SOX9 expression. Once enough SOX9 proteins accumulated, they started to take things into

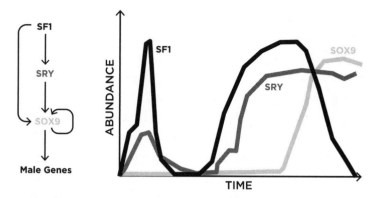

Figure 7.4: The feed-forward loop (left) and its time-dependent function (right). A brief pulse of Sf1 protein expression is not sufficient to turn on *SOX9*, because not enough SRY can accumulate. In contrast, a long pulse of Sf1 causes the accumulation of SRY, and together the two switch on *SOX9*.

their own hands. SOX9 proteins went back to their own gene (these proteins are transcription factors) and further increased the production of their own kind, making sure their levels remain high throughout your life. But beta-catenin had no chance to rise to critical levels. It was vastly outnumbered by SOX9, and its suicide mission of binding to SOX9 led to its removal from what were to become your testicles. SOX9 was free to reign, irreversibly pushing your cells' fate onto the path to testicles.

It is no coincidence that the SOX9 protein manages its own expression, creating a positive feedback loop. This guarantees that once *SOX9* is switched on—meaning it has accumulated beyond a certain threshold—it stays that way: development into testicles is a one-way street. There is another interesting detail in this computation, called a "feed-forward loop" (Figure 7.4). *Sf1* manages SOX9 levels, but it does so with the help of another protein, SRY. As it turns out, SRY, too, is turned on by *Sf1*. Why this complicated structure? Why doesn't *Sf1* manage SOX9 by itself? Why does it first induce the production of a second manager to help with this task? The second manager seems to be a safeguard against instability. Management in the society of genes is by no means perfect. If, by accident, *Sf1* was expressed for a brief moment in the female embryo, and if that alone was enough to turn on SOX9, testicles might accidentally form within a female body. But that cannot happen due to the construction of the feed-forward loop: only when Sf1 proteins have been around for long enough will sufficient SRY have accumulated to help *Sf1* turn on SOX9. In this way, the management makes sure that small inaccuracies cannot disturb the set course.

Because stability in the presence of brief, accidental variation is important in many systems, such feed-forward circuits are found again and again in your genome. Similarly, positive feedback loops (which keep a system turned on once it has been started) and negative feedback loops (which stop the production of a transcription factor once enough copies have been made) are frequently useful, which makes them an important part of the genome's management structure.

So far, we have looked at only one mechanism of regulation, or one type of computational circuit: transcription factors that bind to your genome, inducing or repressing genes. But the "computer" running inside each of your body's cells is even more complex than that. Evolution is a tinkerer, so whatever can get a calculation done is employed. Computations are also done through proteins and RNA interfering with transcription and translation, through the destruction and stabilization of messenger RNAs and proteins, and through chemical modifications that shut down or open up whole sections of your genome.

Master Regulators and Hopeful Monsters

We have seen that even a small number of different parts can turn into an astonishing array of possible structures, depending on how they are wired together. This principle lies at the heart of evolutionary change, responsible for far more fundamental changes than a bigger brain or an improved ability to modulate air pressure. For example, one specific mutation in a fly's genome causes two extra legs to sprout from its head instead of the antennae that belong there. How could the change of a single letter be responsible for such horrific consequences? Consider that the originality of the mutation's effect is limited: the fly already has several pairs of legs. This is just another pair, not novel appendages. Moreover, the legs emerge in the spot at which another body extension, the antennae, belongs. The mutation did not invent a new body part—it transformed one part of the body into another.

In 1900, the British geneticist William Bateson published a catalog of such transformations. Among them were men who had an extra set of nipples and those who had an extra set of ribs. Bateson concluded that changes in nature are often discontinuous, meaning they occur in jumps, which contradicted Darwin's idea that evolution is a gradual process. While Darwin was right that in most cases evolution occurs gradually, there is no law against the occasional jump. Gradual changes are more popular in the history of the society of genes simply because those are more likely not to upset the survival machines they encode. The transformations documented by Bateson, though, provided overwhelming evidence that evolution can occur in jumps.

Such transformations can be horrific or ridiculous, but sometimes they produce what are called "hopeful monsters," individuals of increased fitness. Consider, for example, the mutation that transforms a two-winged fly into a four-winged fly. The extra pair of wings sprouts in the same place that normal flies have small appendages called "halteres," which are used for balancing. For a flying insect, four wings may be better than two wings, at least under certain conditions. Many other insects, including butterflies, indeed have two pairs of wings.

But how can a single mutation orchestrate an entire appendage, complete with many different, specialized types of cells, all masterfully organized? This fantastic mutation debilitates a gene called *ultrabithorax* (or *Ubx* for short). *Ubx* is responsible for managing the production of the halteres, which are located in the same place where wings first developed in an ancient fly ancestor—in other words, in the fruit fly, the second pair of wings evolved into halteres.

Butterflies, which have retained their original two pairs of wings and have no halteres, also have the *Ubx* gene. In the butterfly, the second pair of wings differs from the other in the size and shading of eyespots. *Ubx* controls exactly what to develop in that particular segment in all insects.

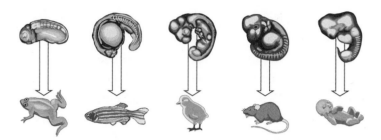

Figure 7.5: Despite the vastly different body plans of animals (bottom), their embryos look surprisingly similar at the phylotypic stage of development (top).

Ubx is a senior manager, a transcription factor that controls an entire orchestrated program. The difference between *Ubx* in flies and in butterflies is which sets of genes are managed: in flies, *Ubx* turns on haltere-making genes, while in butterflies, the same manager controls the genes needed for a second pair of wings.

Studying the gene managers of embryonic development has led to even deeper insights. Legend has it that two centuries ago, the biologist Karl Ernst von Baer was faced with an interesting predicament. The labels on his vials containing embryos of reptiles, birds, and fish were so worn down as to be illegible. So the world's greatest embryologist tried to identify the embryos by eye and found he couldn't do it! He had discovered that at a certain embryonic stage, all vertebrates look essentially the same.

This particular time in embryonic development is called the "phylotypic" stage: this is when the embryo begins to assume recognizable features typical of vertebrates (Figure 7.5). The phylotypic stage represents a general layout on which specialized features—such as the turtle's shell, the pig's snout, or your large brain—can be mounted later in development. In *The Origin of Species*, Darwin invoked von Baer's observations as further evidence that species are related to one another by common descent.

What does this teach us about how organisms are built? It took more than one hundred years to figure out why animals are so similar at a particular embryonic stage. For this we have to first return to the fruit fly. Two unexpected facts were discovered about the handful of genes in flies that, if mutated, can transform whole body parts (antennae into legs, or halteres into wings). First, when the transforming genes were mapped to the fly genome, they were found to be neighbors. Second, during development, these genes were turned on in the same order in which they are written on the chromosome. It is as though this region contains a master plan for constructing the fly.

But the most revealing discovery occurred when this region in the fly was compared to the same region in other animals. Up until the 1980s, it was assumed that different animals had very different sets of genes. It was thus completely surprising to find that the transforming genes from the fly—together called the *HOX* family—have counterparts in the worm *C. elegans* and in the mouse, where they occur together in an almost identical arrangement.

The *HOX* genes of different animals not only have very similar letter sequences, but they can also fill in for one another. If a worm or a mouse has a defective copy of one *HOX* gene, its embryonic development can be rescued by providing its genome with the corresponding fly gene. Most animals have the *HOX* gene cluster (comb jellies, for example, do not). Given that *HOX* genes manage the identity of body parts, we now have an idea of why there is a stage when embryos of different animals look alike: they have the same senior management, the same set of *HOX* genes.

The genomes that control the development of different animals are surprisingly similar to one another, even aside from sharing *HOX* genes. For example, the same three crucial manager genes oversee muscle development in all animals, but with results as different as a mouse and a fly. It is not the genes themselves that differ, but the ways in which they interact with one another. The networks of managerial interactions are intricate webs of cooperation and obstruction. Some of the interactions are the same across all animals, but some have completely changed through evolution.

After this tour through animal development, let us look at a completely different kind of development. When times are rough, the bacterium *Bacillus subtilis* can seal its offspring into a sort of time capsule from which it will emerge when things improve. When nutrients run low, *B. subtilis* begins to build what is called a "spore," a specific cell type that is so resistant that it can survive boiling water and radiation from an atomic bomb. When the spore is complete, the mother cell commits suicide, leaving its child behind to wait for better times, even if it takes 1,000 years, at which point the spore resumes the business of a normal *B. subtilis* cell. Again, a network of managers is responsible for controlling this process. It starts with a senior manager called *Spo0A*, which turns on a whole cascade of other managers, until the entire cell commits itself to the making of a spore. This is much like the way some cells in your body commit themselves to making an ovary or testis during embryonic development. It turns out that *Spo0A* is no stranger: it is the bacteria's *HOX* gene by another name. Across all cellular life, the most-senior managers are distant cousins, performing the same type of job in very different organizations.

As a final thought on gene regulation, consider what happens if you lose a finger through an accident. Why don't you just grow a new one? You built that finger in the first place, so your cells know how to do it. Why can't they do it again? While most of us are lucky enough to keep our fingers intact, the same cannot be said about our teeth. Wouldn't it be nice if we could replace teeth with organically grown new ones?

Over the past sixty years, we have learned to read and understand much of the genome's language, but we know very little about how to speak this language ourselves. Will we one day have learned how to change genetic programs, helping our bodies to replace defective body parts? Salamanders can regrow limbs or eyes, and other animals have similar abilities to regenerate lost body parts. Eventually, by studying the differences between their genomes and our own, we, too, may be able to talk our cells into regenerating parts of our bodies.

Gene regulation allows for a wide range of possible phenotypes from the same set of genes. But not all novelty is the result of mixing and matching. Sometimes, new members must be introduced into the society of genes.

8

Theft, Imitation, and the Roots of Innovation

Originality is nothing but judicious imitation.

—Voltaire

IN THE EARLY TWENTIETH CENTURY, TWO FORMS OF COLLECTIVE societies arose in the place that was to become the state of Israel: the kibbutz and the moshav. Imagine two young women (let's call them Ada and Eve), one coming of age in a kibbutz, the other in a moshav. What career paths would have been open to them?

In a kibbutz, of which there were hundreds, a system of collective living was set up to put into practice a kind of utopian socialism. Ada would have grown up together with the other kibbutz children, separated from her parents. The families shared all resources; there was a single bank account for the entire collective. As she grew up, Ada would have been expected to stay in the kibbutz and marry one of her playfellows. She would lend her hand to the main business venture of the kibbutz, which might be a factory specializing in diamond cutting or in manufacturing drip irrigation. As each generation led to an overall increase in the number of kibbutz members, new jobs would have had to be created. To meet this need, an existing job often would be divided into several specialized ones. For instance, if Ada's mother performed two related tasks in the kibbutz's factory—say, polishing the cut diamonds and inspecting their quality—these tasks might have been divided between her two daughters. Ada would specialize in quality inspection and her sister in polishing. This specialization would allow Ada and her sister to become higher-level specialists than their mother had been.

Growing up in a moshav, Eve would have had a different experience. A moshav is a community of cooperative farmers; many moshavs are still in existence today. The farms are of a fixed size, with each family producing a specific crop or product. The goal is for the community to be self-sufficient. For each family's farm to remain intact, only one child inherits the whole plot. That child is bound to continue farming. The rest of the offspring get no land at all. Let's say that Eve grew up in a family that specialized in making goat cheese. She did not inherit the property, so she had to consider other career options. She could take her experience in making goat cheese and move to a new community, perhaps another moshav, one that did not have a cheese maker. In such a transfer of skills, the moshav would have goat cheese, and Eve's future would be secured: both Eve and her new community would benefit.

An Eye from an Eye

The situations of kibbutz and moshav children are representative of the different ways in which new members of a society can make a living: by specializing and forming a new trade, or by transferring their skills to another society. Similarly, specialization and skill transfer are two major ways new members are integrated into the society of genes.

As a first example, consider how in our ancient animal history, new genes were recruited to allow our predecessors to see in color instead of monochrome. The colors we see derive from signals received by just three types of receptors in our eyes, each tuned to a restricted range of light frequencies: one each for red, green, and blue. A separate gene produces each of these receptors. When light comes into the eye, it thus triggers three specific signals that are then processed by the brain to let us distinguish millions of colors.

To understand how this system of three receptors evolved, it helps to first consider a more recent process: the invention of color television. For our perception of virtual reality, this was a dramatic step. When television was still black and white, many people believed that they dreamed in black and white. But technologically speaking, how difficult is it to display colors once you can already display black and white?

Color television, able to display millions of colors, works by tricking the human color-vision system. Because three receptors are all we have to detect color, color television needs to provide only one signal for each of the receptors. Purple light, for example, equally stimulates your receptors for red and blue. As your brain receives these signals, it will register the origin of the signal as purple. To make a television set capable of projecting colored pictures, the inventors triplicated the light-projecting system already in place for black-and-white televisions and gave each subsystem a different color: red, green, or blue. By copying the parts of black-and-white television and

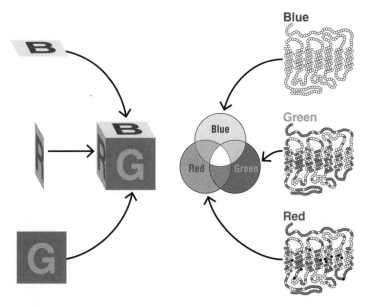

Figure 8.1: On the left is a representation of how color television displays an image. Each pixel displays three different spots in blue, red, and green; our brain pieces them together into a colored image. On the right are the opsin light receptors that together can capture many colors. Amino acids that differ between the blue and green proteins are indicated with filled gray circles. Changes in the red opsin relative to the green opsin are indicated with black.

adding a small "mutation" to each of the copies, color television was born (Figure 8.1, left).

In a similar way, color vision was invented by the society of genes. Monochrome vision (essentially, black and white), with a single type of light receptor, already existed in much earlier animals. This receptor was duplicated and modified repeatedly, resulting in the three light receptors with different color preferences now found in our society of genes. We know that color vision arose in this way because when we compare the genes encoding our three receptor proteins, or opsins, we find that they are very similar in sequence. Such highly similar sequences are so unlikely to have arisen by chance that they almost certainly stem from a common ancestor (Figure 8.1, right).

Gene duplication is a special type of mutation. Such mutations can occur through mistakes in DNA replication, when the polymerase DNA-copying machine slips on its template and rereads a part it had already copied. Another frequent accident leading to duplications occurs in recombination during meiosis, the crucial prelude to sex that mixes the two matching chromosomes an individual inherited from his or her parents (Chapter 3). When a region on one chromosome is accidentally lined up with the wrong region on the matching chromosome, everything in between the two regions will be deleted from one of the resulting copies, while it will be duplicated on the other copy.

Early in the history of opsins, our distant animal ancestors had just a single opsin gene. Four consecutive duplications of this gene resulted in five opsin genes spread out across the genome. One of the five is a rod opsin, unable to distinguish color but highly susceptible to low light levels. That is why all cats appear gray at night—only the rod opsins are sensitive enough to see them at all. If this rod opsin were all we had, black-and-white TV would be all we'd ever need. The other opsins—the cone opsins—allowed our ancestors to distinguish color. Our ancestors' color vision became richer with each additional duplication of an opsin gene and its subsequent modification through mutations.

At the time of dinosaurs, the first mammals were nocturnal creatures, so they had little need for color vision, which relies on high light levels. They lost two of the four cone opsins that had evolved in their ancestors. Thus, most mammals see a world of color that is much poorer than yours, unless you are color blind. In that case, your genome, like most mammals, contains only two functional cone opsins, and you see color about the same way they do.

More recent ancestors of apes and monkeys found color vision to be advantageous, because they were not nocturnal, and seeing colors helped them forage in bright daylight. Through another chance duplication of one of their cone opsin genes, brought to prominence in the society by natural selection, they obtained the ability to distinguish three basic colors, giving them trichromatic vision. A richer color vision might have enhanced their ability to detect fruit. Furthermore, it may not be by coincidence that humans and other apes with trichromatic vision have lost parts of their facial hair, allowing subtle changes in skin color to be visible in the faces of rivals or mates.

The duplication of a gene, which inserts a second copy into another location on the genome, solves the conceptual problem involved in creating something new. Imagine that mutations change one or more letters in a gene, such that by chance, the mutant version now has a new and useful function. Before it mutated, the gene most likely performed a useful service to the society of genes. How would that old function continue to be performed? If the gene is duplicated before the mutations happen, then one copy can retain the old function. In the words of the biologist Susumu Ohno, who in 1970 first appreciated the importance of this insight: "Natural selection merely modified, while redundancy created." According to Ohno, almost all novelty in the society of genes arises from the duplication of an existing gene. Natural selection will make sure that one copy maintains the original function, while the other gene is free to adopt new functions, ones that may then themselves be promoted by natural selection.

Two duplicates of an opsin gene will not survive if they continue to carry out identical functions, because a random mutation that wiped out one of them would not be penalized by natural selection. If, in contrast, they each specialize in a separate wavelength, and together they perform better than a single nonspecialist opsin did, then the two copies will likely survive and establish their place in the society.

For comparison, consider a saleswoman with two daughters living in an expanding Israeli kibbutz. One daughter might follow in her mother's footsteps, exchanging letters with the kibbutz's existing customers and potential new customers. The second daughter would then be free to explore new opportunities. She might end up setting up a website, adding an Internet distribution channel to the kibbutz's business.

Typically, a newly duplicated gene is identical to its template, and therefore, it is highly similar in function. But duplicates can also acquire radically different properties. Vision, for example, is not only a matter of photoreceptors. The eye needs a lens to focus light onto the receptor, just as a camera needs a lens to focus light beams onto its light sensor. In the bodies of animals, such lenses are formed by a dense solution of transparent proteins called crystallins. The crystallins' main task is to transparently fill up the space of the lens while increasing its refractive index. From their DNA sequence, many of the crystallins are easily recognized as duplicates of genes involved in metabolism; the original copy of one human crystallin specializes in breaking down alcohol.

Animals often do not even bother to first duplicate the gene whose product is used as a crystallin. A protein that makes up one-tenth of the crystallins in duck lenses doubles as an enzyme that helps break down lactic acid, a substance your muscles produce during strenuous exercise. This sort of moonlighting is not uncommon among enzymes; many fulfill several functions that may or may not be similar to one another.

Such multifunctionalism actually provides a fairly gradual mode of evolution for a new gene function: if a job needs to be done, any gene hanging around with at least rudimentary capabilities can be recruited. Further random mutations (or existing variation among alleles) will then be employed by natural selection to optimize the expression and letter sequence of the gene for the added task.

Often, however, there are trade-offs between the two functions, in the same way that it would be rare to find one craftsperson who is equally good at building ships and at making musical instruments. In the wake of a chance duplication of the gene encoding the moonlighting proteins, evolution may seize the opportunity to divide the job and create two specialists.

Does our sense of smell rely on a system similar to that of color vision? We can detect millions of different odors, each one a unique combination of molecules floating in the air. To see millions of colors, all we need is for three light detectors to combine and pinpoint the position of a light signal in the continuous spectrum of wavelengths. But molecules are discrete, and there seems to be no efficient way to distinguish them with just a handful of receptors. Still, a few hundred scent detectors are enough to distinguish millions of scents. Each particular odorant activates a specific combination of receptors in your nose. A specialized part of your brain processes the hundreds of signals, integrating the specific combination of triggered receptors into a scent.

In principle, a multitude of receptors could be generated by assembling new combinations of receptor parts, similar to the Mixies-like strategy employed by the immune system to make millions of detectors for intruder proteins from a handful of preexisting gene parts (Chapter 2). In the case of smell, however, evolution occurred by a simpler route, encoding each of the hundreds of smell receptors in a different gene.

In an early fish-like animal, a first smell receptor was able to recognize one class of molecules. Through a genetic accident, this gene was duplicated in the genome of one individual, where it now occurred in two separate copies. One of the copies may have had a chance mutation, resulting in a slight alteration in which molecule it could smell. Those who inherited the functionally different duplicates might have been better at distinguishing between wholesome and dangerous foods, and so natural selection favored the duplicate smell-detector gene.

This process repeated itself, with the duplicates duplicating again and again, until the original receptor's descendants eventually made up the full smell-receptor repertoire that populates our genomes today. It's like a kibbutz that started with a handful of individuals; they had to perform all tasks needed in the factory. With the increasing expansion of the founders' families, the kibbutz developed into an entire society with an intricate system of specialized labor, split among the descendants.

If you did a census of your genome, you would find that 5 percent of your genes are duplicated smell receptors. In terms of the workforce, detecting scents is the largest business in the society of genes: it is the biggest gene family in the human genome. However, out of the nearly 1,000 smell receptor genes, two-thirds are actually broken. These dead genes (known as "pseudogenes") contain mutations that make them unable to perform any useful function.

Why would your genome carry a graveyard for smell-receptor genes? Why did they die in the first place? Normally, when a mutation occurs that cripples the function of a gene, that mutation does not do very well on evolutionary timescales. The mutated allele is lost from the society of genes, because it puts the individuals who carry it at a disadvantage. This process is called "negative selection," the flip side of the positive (or Darwinian) selection discussed throughout this book, the process that makes a mutation increasingly popular if it enhances fitness.

With the rise of trichromatic color vision in our primate ancestors, it is likely that we came to rely much more on vision than on our sense of smell. Hence, the smell detection system became less important. In such a scenario, a mutation in one smell receptor would not put a person at enough of a disadvantage to seriously influence her survival, and half of her offspring would inherit the now-dead smell receptor without noticing any effects. Over time, many genes with reduced utility have died in such a manner, and many are still in the process of dying. For a given smell receptor gene, some humans may carry a functional allele, while others have an allele that is dead. Which version will eventually remain in the society of genes may be just a matter of chance.

Dogs have only two different types of color receptors (cone opsins) and thus perceive colors the way a color-blind human does. Dogs' poor color perception is more than compensated for, though, by their amazing ability to detect and distinguish smells. Indeed, dogs have about the same number of smell detectors in their society of genes as we have in ours, but nearly all of them are still alive and kicking.

All in the Family

In any complex society of genes, gene duplications are very common. While our own genome does also contain genes that occur in just a single copy, gene duplicates account for the lion's share of our genes. Families of genes that result from several rounds of duplications come in many sizes. As we have seen, the largest family is the scent receptors, which are roughly 1,000 genes strong, while the family of opsins for vision is quite small.

But what is a family, anyway? Your own nuclear family is part of a larger family that includes grandparents and an even larger one encompassing cousins several times removed. You can also view the scent receptors as part of a giant family that includes gene cousins involved in communication among cells. The smell receptors and their cell-to-cell communicator cousins were each founded by duplicates of the same "grandmother" gene (Figure 8.2).

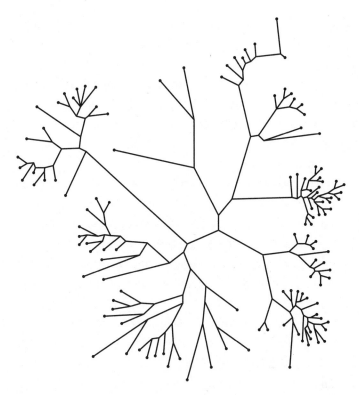

Figure 8.2: A gene family tree. The circles indicate genes, and the lines show their relationships. While all genes belong to the same family, it is difficult to decide how many separate nuclear families are represented—in this diagram, is it one, four, or eleven?

Ancient duplicates have often mutated beyond our ability to recognize them today, just as most human families can trace their distant relatives no more than a few generations back. If we take this to its logical conclusion, it becomes plausible that almost all of your genes are members of one big family, including those genes we thought are present in just a single copy. Their ancestry might be traced back to a time when there were just a handful of genes. Through a long series of duplications and modifications, that handful eventually turned into humanity's rich society of genes.

Duplications can occur on any scale. A duplication may encompass only a few letters, lengthening a gene. Or it might cover entire regions of the chromosome, affecting many genes. A whole duplicate chromosome may even be inserted into the cell by mistakes in cell division.

The mother of all duplications, however, would be a duplication of the entire genome. The cellular machinery is optimized for dealing with two sets of matching chromosomes, not four, and thus the chance for survival of an animal embryo that inherited such a radical change is small. Even if a duplicated genome could build and control a viable individual, this individual would not be able to produce healthy offspring with a mate that had the normal two copies of each chromosome. Their children, who would inherit half of each parent's genome, would have three copies of each chromosome. They would be sterile, because when it came time for them to make egg or sperm cells, it would be impossible to properly divide these copies in half. Despite these major obstacles, however, whole-genome duplications are successful once in a long while. Our own human society of genes is the descendant of not one, but two complete duplications that happened in our fish ancestors about 400 million years ago.

These genome duplications left a big mark on the society. One good example is the *HOX* gene family, the top managers of the society of genes. As we saw in Chapter 7, *HOX* genes set up the body plans of animals by controlling where and when other genes are switched on in the developing embryo. Worms and flies have only a single cluster of *HOX* genes on one of their chromosomes (in flies, it is split into two parts), but your genome has four such clusters on four different chromosomes, which is just what you would expect after two successive rounds of whole-genome duplications. With more and more specialized managers of body construction, the corresponding societies could specify more complex bodies. The increased complexity of body plans seen in the vertebrates may indeed have its root in this particular, unusual arrangement of *HOX* genes. Take your thumbs, for example. All other fingers express three genes from one of the *HOX* clusters, but these three are not active in your thumbs, which explains why they have a different shape from your other fingers.

Genome duplications are not restricted to our own history; they have occurred in plants, in fungi, and in fish. Genome duplications are giant leaps that occur in a society of genes from time to time, leaps that are clearly at odds with Darwin's conception of gradual evolutionary change. Gradual changes do appear to account for most changes in the society, but the very few genome duplications that do occur have dramatic consequences.

What does a genome duplication mean for the society of genes? When we view genes as a society, each gene is an industry in which multiple alleles compete. When a gene is duplicated, it is as though an entire industry has been duplicated. A full-genome duplication, then, is a situation in which all industries have been copied. In such a duplicated society of genes, many of the additional genes are redundant, the equivalent of having two entire industries devoted to baking, repairing cars, and so forth, when one of each would suffice. Many duplicated industries do not survive for long in the genome: random mutations that eliminate a redundant gene's function go unnoticed by natural selection. A duplicate's only chance for long-term survival lies in specialization, in the same way an all-purpose bakery might split up into three business, one specializing in bread, another in bagels, and the third in donuts. A duplicate has a limited time, a kind of probationary period, in which to take on such new roles—only then it will be protected by natural selection from chance debilitating mutations.

Hemoglobin, a protein assembly inside your red blood cells that carries oxygen to the furnaces of the cells, is a nice example of how duplicates make themselves useful by becoming specialists. Your hemoglobin is built from alpha globin and beta globin proteins, encoded by two different genes in your genome. These two genes are so similar that they are clearly duplicates of an ancient, nonspecialized globin gene, which must have once formed functional hemoglobin all by itself.

Your genome actually contains even more duplicated copies of globin genes with slightly different properties, specialized for specific situations. One of them, gamma hemoglobin, is currently not used by you at all. It was put to work only when you were a fetus, that is, from about six weeks after your father's genome half had been delivered to your mother's genome half in the egg cell. Since birth, though, your hemoglobin consisted mostly of assemblies combining two alpha and two beta globins. The strength with which hemoglobin binds oxygen differs between these hemoglobin types. The

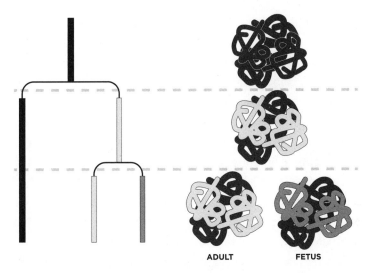

ADULT   FETUS

Figure 8.3: Most adult hemoglobin is composed of two alpha globins (black) and two beta globins (light gray). In fetal hemoglobin, the beta globins are replaced by gamma globins (dark gray). The family tree to the left shows the relationships among these three hemoglobins: an ancestral globin gene (top) was duplicated, leading to specialized alpha and beta globins (middle). The beta globin then duplicated again to give rise to an additional gamma globin specialized for the environment inside the uterus.

fetal hemoglobin binds oxygen much more avidly than does the adult hemoglobin and is thus more efficient at sucking oxygen out of your mother's blood stream through the placenta (Figure 8.3).

The Society's Lego Set

If kids and parents agree on something, it might be that the Danish construction set Lego is the greatest toy ever invented. Nowadays, to motivate repeat purchases, Lego sets come with pieces tailor-made to construct one particular thing, such as a tow truck or a Death Star. The original idea behind Lego was much simpler and much more visionary. The design of its rectangular blocks allowed you to build almost anything imaginable from a sufficiently large number of pieces. Many genes are built with a similarly efficient modular construction system. These genes are constructed by combining what we call domains, simple building blocks that have been duplicated in the society of genes again and again.

Take the genome's managers, for example. In Chapter 7 we saw that transcription factors are proteins that bind to the regulatory switches at the start of other genes, thereby controlling where and when these switches are turned on and off. The *FOXP2* gene involved in your speech and in birdsong contains two types of such domains. One is a winged helix region, named after the butterflylike shape of the protein section it encodes. This domain has just the right shape to straddle a matching stretch of your genome's DNA between its wings. The second unit is a zipper domain, and its role is to bind to the zipper domain of other *FOXP2* proteins (Figure 8.4).

Several thousand genomic Lego-like domains have been identified. Each domain typically performs one specific function. More than 80 percent of your genes contain at least two different domains, and these domain combinations make them into specific and complex machines. An almost endless variety of novel genes can be created through new combinations of these domains. This is similar to your immune system's strategy of creating a vast diversity of antibodies by rearranging the variable, diverse, and joining (VDJ) regions of the genome (Chapter 2). An important difference is that the domain rearrangements are not fostered by a specialized machinery on a

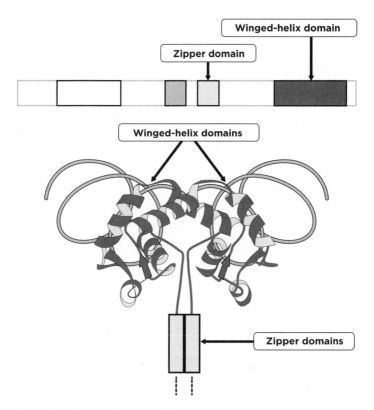

Figure 8.4: The *FOXP2* gene and its zipper and winged-helix domains. The top image, a schematic of the gene sequence, shows the domains as shaded blocks. The bottom image shows two FOXP2 proteins held together by their zipper domains. The winged-helix domains are the dark and light gray spirals straddling a matching piece of a chromosome (a control element to be bound by *FOXP2*).

routine basis but are rare genomic accidents. If two otherwise different proteins contain one domain of the same type, at least one of them was most likely assembled by accidentally mixing parts of earlier genes—a shuffling of the domains. Thus, new genes typically originate as duplicates of other genes or from a remixing of duplicated parts of existing genes.

The Export/Import Business

Both as hunters and as prey, early humans would have benefited from being able to fly, so why didn't your ancestral society of genes copy the relevant genes from a bird?

First, it is highly unlikely that adding one, or even a few, bird genes to your genome would result in a humanoid that can fly. Furthermore, the copied bird genes could work only if they are integrated into sperm or egg cells (the germ line), but a strong barrier prevents foreign DNA from getting there. Even if a foreign gene sequence made its way into a sperm or egg germ cell, it would be blocked from entering the cellular compartment that houses the chromosomes. These substantial barriers are probably the result of a trade-off. Integrating the good stuff from other organisms may sound nice, but most sequences that would line up for integration into your genome are not beneficial.

Those restrictions apply to all complex animals and plants. But the story is different for bacteria, including the humble *E. coli*. Bacteria have several means for picking up foreign DNA: they can swallow it for food, or visiting viruses can accidentally bring it along. Bacteria do not have a separate germ line with specialized sperm or egg cells: they are just one cell that keeps dividing. Their genome is freely accessible in the cell's interior. What is equally important is that bacteria have less to lose than a multicellular organism. If a bacterial cell tries out something dangerous, like integrating a piece of DNA from a complete stranger, it may die. But a bacterial organism consists of just one single cell, and its genes have many identical siblings in the surviving sister cells. If any newly integrated

DNA is even slightly detrimental, its carrier gets pushed aside by its numerous twins. Since bacterial populations tend to be vast, the loss of just one cell is a cost easily borne. However, there are rare cases in which the integrated DNA provides an unexpected advantage. When that happens, the descendants of the lucky carrier can take over the whole bacterial population.

The ability to integrate foreigners into their societies of genes gives bacteria an enormous evolutionary flexibility. If a bacterium finds itself in a new environment, it is likely to encounter other bacteria that have already adjusted to life there. Each new arrival can speed up its adjustment by picking up DNA from those acclimated residents. This process is rather similar to that undertaken by our ancestors when they acquired Neanderthal immune genes in the hostile environment outside Africa; however, that transfer of genes involved the act of sex, so it was restricted to individuals belonging to the same species.

It's important to note that what bacteria steal from other bacteria is DNA, not proteins. By picking up DNA, bacteria are engaging in a form of intellectual theft. Who benefits when a bacterium evolves an innovation, such as resistance to an antibiotic, and passes it around to other bacteria? Surely the bacterium that invented it and the bacterium that stole a copy: they get a better chance to survive. But the real beneficiary is the gene that imparted the antibiotic resistance. Because that gene is not restricted to the genome in which it first arose, and because different bacteria regularly exchange DNA, the gene with that particular antibiotic-resistant innovation can be passed around until, eventually, it provides a wide array of bacterial species with the ability to survive the antibiotic. In this way, the resistance gene establishes itself across several different societies of genes.

Resistance to drugs by copying the genes conferring resistance has been well documented. Penicillin and several other antibiotics have already ceased to work against many of our bacterial enemies. As soon as one species of bacteria figures out a way to evade the drug, many other species copy the trick. The discovery of antibiotics was a giant leap in our ability to defend ourselves; at the same time, it was just one episode in the long and complex relationship between animals and bacteria. Humanity has science on its side, while disease-causing bacteria, with access to one another's genes, have strength in their unity.

In our guts, bacteria find an almost ideal environment for the exchange of antibiotic-resistant genes. Your intestines are home to an incredibly rich community of about 100 trillion microbes from several hundred different species. These microbes often form biofilms, sheets consisting of closely connected cells from different bacteria. The close contact between cells greatly enhances the opportunity for DNA transfers. Well over 90 percent of humans in developed nations carry antibiotic-resistant bacteria in their guts. The residents of our intestines thus act as a reservoir for antibiotic-resistance genes, which can be passed on to other bacteria that are just traveling through.

The genes that jump among bacteria are, of course, not restricted to those conferring antibiotic resistance. Typically, bacterial societies of genes transfer genes involved in the bacterium's interaction with the environment. These genes often encode transporters or enzymes for the processing of nutrients. The evolutionary action is at this interface: as soon as the environment contains a previously unknown food source, a bacterium first needs a transporter to get that food inside itself and then an enzyme to digest it. If genes of other bacteria in the neighborhood have already evolved the relevant transporters and enzymes, then intellectual theft is the obvious solution. If you examined the genome of the *E. coli* bacteria in your gut, you would find that more than a third of their nutrient transporters were copied from other bacteria over the course of the past 100 million years.

Intellectual theft of genes, called horizontal gene transfer, can be viewed as being a more efficient duplication system than the one that makes copies of genes within the society. If the same gene finds itself in two bacteria that are related but have to adapt to different environments, the two copies may evolve in different directions. If a later horizontal gene transfer then puts them back together into the same genome, what results resembles a gene duplication with already-diversified copies. The horizontal transfer of a gene can be viewed as gene duplication at the scale of a whole ecosystem of bacteria.

Human genes can mingle only with those from their own society of genes, but bacterial societies can in principle draw new genes from a universal gene pool shared by all bacteria. Even so, bacteria are unlikely to meet comrades that live in very different environments, and they are more likely to benefit from genes that have been proven in their own environment and in species not too unlike themselves.

This kind of genetic intellectual theft is analogous to the situation of an expanding family in an Israeli moshav, with children fanning out to offer their skills to other moshavs. From the perspective of the adopting moshav society, such transfer is clearly advantageous, leading to new and previously inaccessible avenues of growth.

Copying and intellectual theft in the society of genes are the main mechanisms for integrating new genes. For the most part, these changes are gradual, but in rare cases the effects can be spectacular. When an entire genome is duplicated, whole new avenues of functionality open up. As we shall see in Chapter 9, intellectual theft of a whole genome, though rare, is also possible, and has results of even grander proportions.

A Secret Life in the Shadows

In union there is strength.

—Aesop

THINK OF SOMETHING THAT YOU ARE GOOD AT, AND ASK YOURSELF
why you are good at it. You may think you know the answer, but
maybe you're wrong. That's what happened to Jim Bodman,
the chairman of the Vienna Sausage Company in Chicago. He
told the following story on the radio show *This American
Life*. Jim thought he knew how to make good sausages. They
were a great success, after all. He had the detailed recipe,
with all of the specifics about the spices, the ovens, the water,
and the temperatures. But he really didn't have a clue, as he
found out after he moved his company to a new, state-of-
the-art facility on the north side of Chicago in 1970. In the
new place, the sausages just didn't have the same bite; even
the color was wrong. After a year of checking all possible reasons they could think of, his team still couldn't figure it out.

In the old plant, there had been a guy named Irving. Everybody had loved him, but he had chosen not to move with
the company to the new location. Irving's job had been to
take the uncooked hotdogs from the freezer to the smokehouse. The old plant had grown in a haphazard way and was

not specifically built for the tasks at hand. And so Irving had to take a 30-minute trip with the hotdogs, wandering through a maze of hallways, right through the areas where corned beef was cooked, before getting into the smokehouse. This trip had the unintentional effect of warming up the hotdogs before they arrived at the smokehouse. In the new plant, they were moved from the freezer to the smoker in seconds.

Irving's trip turned out to be the secret ingredient! Once Jim and his team realized this, they built a whole new room to simulate Irving's path, and the hotdogs got their old bite back. During all those years at the old plant, a secret ingredient had been hiding in the shadows.

Throughout this book, we have been telling you how your genome controls your body and your life, from diseases to sex. But just as with Jim Bodman's sausage factory, there is a secret ingredient in your cells, something beyond what is explainable by looking at your forty-six chromosomes. Our story starts with the observation that there are two fundamentally and vastly different ways for a cell to make a living: as part of a conglomerate or alone. The bacterial cells in your gut go through life one by one as single cells: they are part of a community, they cooperate or quarrel with their neighbors, but their fates are not inextricably linked. Human cells, in contrast, are part of a giant corporation, in which all cells are completely dependent on one another: only the sum of your cells can call itself an individual.

Animals are conglomerates of up to trillions of cooperating cells, each with a plan for the organism's division of labor. There is astounding variation in how this cooperation is organized: think of sponges, jellyfish, snails, worms, flies, starfish, and frogs. And plants, fungi, many algae, and slime molds are made up of cell conglomerates, too. While there are untold different kinds of bacteria, none of them can join forces to build anything as large and complex as an animal or a plant. Why not? What's stopping them?

One reason is size. Multicellular organisms are bigger not only because they consist of more cells, but also because each of your cells is larger than a bacterium. And not only slightly larger—in terms of volume, your cells are about 1,000 times larger than an *E. coli* bacterium. Your cells are bigger because each one of them needs to contain all the instructions for building and controlling the specialized structures needed by your complex organism (Figure 9.1). A much larger genome needs to be packed into each cell, and your genome, too, is about 1,000 times bigger than that of a bacterium. A

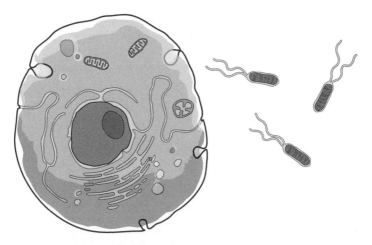

Figure 9.1: The cells of animals (left) are much bigger than those of bacteria (right) and require proportionally more energy.

larger cell size is also required for the specific function of many cells: your brain, for example, works on principles that require cells of particular shapes and dimensions, which could not be built within the limits of bacterial cell sizes. The same goes for the cells in your muscles, your blood, and your immune system: their size is highly relevant to their function.

It turns out that bacteria cannot produce enough energy to run much bigger cells. What allowed us to make cells that big? There is something lurking in the shadows of our complex cells; something that explains more than just cell size. We told you that your genome has forty-six chromosomes. We lied. We depicted evolution as a process driven only by genomic mutations, but that's not exactly true either. The reality is more interesting. At the heart of these mysteries is an ancient merger of businesses.

The Birth of a Kingdom

To understand what enabled our complexity to evolve, we must trace backward into deep evolutionary time. Following Darwin's ideas, we have implied that all life, from humans to bacteria, is part of one large family tree. A representation of evolution in the form of a tree is the single illustration in Darwin's magnum opus, *On the Origin of Species*, perhaps an indication of how strongly he wanted to get this idea across (Figure 9.2). He wrote: "As buds give rise by growth to fresh buds, and these, if vigorous, branch out and overtop on all sides many a feebler branch, so by generation I believe it has been with the great Tree of Life, which fills with its dead and

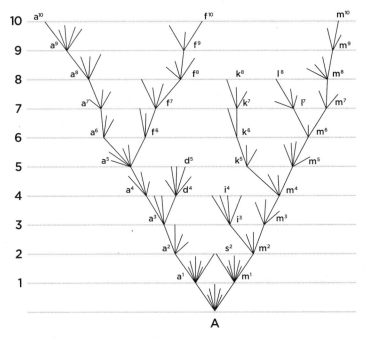

Figure 9.2: This phylogenetic tree forms part of the only illustration in Darwin's *On the Origin of Species.* "A" indicates a given ancestral species. Each horizontal line indicates the passage of 10,000 generations. At each time, the species can spawn new varieties, most of which will not survive. At the top are the remaining living varieties, which will eventually form independent species.

broken branches the crust of the earth, and covers the surface with its ever branching and beautiful ramifications."

Ever since the publication of Darwin's book in 1859, the possibility of revealing the historical narrative of life on earth as a tree has captured the imagination of scientists. Throughout the nineteenth and twentieth centuries, the tree of life underwent many major revisions, and both deep and shallow branches are still constantly updated.

In 1925, Édouard Chatton, a researcher at the Louis Pasteur Institute in Paris, discovered that there are two types of cells: those that have a nucleus—a special room inside the cell that houses the cell's genome—and those that do not. The organisms built from the former cells were called eukaryotes, a term that combines the Greek words for true (eu) and kernel (karyote). All multicellular organisms—animals, plants, and fungi—belong to the eukaryotes. Bacterial cells lack a nucleus; their genome is suspended directly within the confines of the cell. Because these cells were regarded as a more primitive life form, they were given the name prokaryotes; using the Greek prefix pro- to mean "before."

Chatton could not substantiate the notion that having a nucleus, the special compartment for DNA, was evidence of a deep evolutionary divide between bacteria and multicellular organisms. What if organisms from different lineages had invented or lost such cellular rooms multiple times? For the deep relationships of life to be understood, finding evidence to support or refute a major division between prokaryotes (bacteria) and eukaryotes (multicellular organisms) became a pressing issue. What was needed was a way to show that eukaryotes are more closely related to one another than they are to prokaryotes, and vice versa.

But how is it possible to make a family tree that contains organisms as different as humans and bacteria? For the first hundred years following the publication of Darwin's book, evolutionary trees were based on the visible characteristics of species. To make an evolutionary tree for birds, the length, shape, and color of beaks could be examined. Scientific controversies raged over how to interpret changes in physical characteristics, and, consequently over how the species were truly related to one another. Without objective criteria, little could be done to settle these quarrels. Our ability to identify the letter sequences of DNA forever altered this challenge.

In Chapter 4, we built a genomic family tree for a single human family, starting with recent ancestors (like great-grandparents) down to the current generation. Using the same methods, we can construct an evolutionary tree that spans all of life, from bacteria to humans. The way to compare the genomes of such diverse species is to look at individual genes that have such general functions that they have been around from the beginning of life and are still found in all societies of genes. For a particular universal gene, counting the number of differences between copies in two organisms will give a rough measure of the time that has passed since the sequences had their last common ancestor. The gradual accumulation of genetic change over millions and even billions of years means that the more similar any two species are in terms of their DNA, the more closely related they are. For certain age estimates using the DNA method, fossils were available as further evidence. To cross-check the relations, radiometric dating was used to estimate the age of the fossil and, generally, the methods agreed very well.

Using such DNA comparisons, it is possible to reconstruct the full evolutionary tree for a representative gene that is present in all species of interest. An example of such a universal gene is the one responsible for making a molecule called 16S ribosomal RNA, a crucial part of the ribosome, the molecular machinery that glues amino acids together to make proteins. Every cellular organism has this gene, be it human, bacterium, or plant.

This one gene can be used to build a complete tree of life, a feat unthinkable with methods that rely on comparing the organisms' outward appearances.

In the late 1970s, Carl Woese and George Fox used this method to build the first comprehensive evolutionary tree. Their results were totally surprising. Instead of the two groups hypothesized by Chatton—eukaryotes (including multicellular organisms) and prokaryotes (bacteria)—there were three! One group of bacteria, which includes our friend *E. coli*, clearly differed from the eukaryotes; but the big surprise was that Woese and Fox found another group of bacteria within the eukaryote branch. These two types of bacteria were different organisms, as different from each other as we are from bacteria. The group of bacteria that appeared to be more closely related to eukaryotes was named archaebacteria, and the other group of bacteria was renamed eubacteria ("true bacteria"). Simply by comparing letters in the DNA, Woese and Fox had discovered an entirely new domain of life.

What are archaebacteria? Before Woese and Fox's discovery, these cells had not appeared to be fundamentally different from other bacteria. A closer look at their genome revealed that even though archaebacteria are similar in size to eubacteria, many of their genes are profoundly different. One striking difference is the way in which archaebacteria and eubacteria build their cell walls. To carry out their own type of cell-wall construction, each one of these types of bacteria has its own set of genes.

Archaebacteria live for the extremes. They call home some of the toughest places on earth, such as hot springs close to water's boiling point, alkaline and acid waters, the digestive

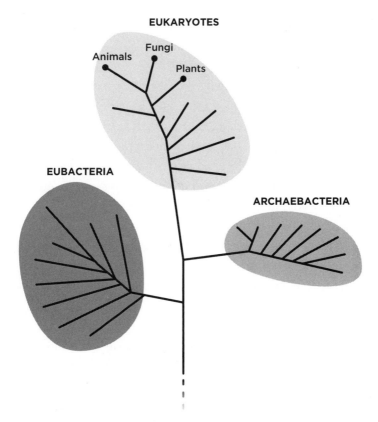

Figure 9.3: The tree of life built by Woese and Fox was based on comparisons of *16S RNA* genes. The deepest divide is between the eubacteria on one side and the archaebacteria and eukaryotes on the other.

tracts of cows, and the bottom of the ocean. Some archaebacteria have even been found living on gasoline.

In Woese and Fox's tree of life, eukaryotes branch off from the lineage of archaebacteria (Figure 9.3). That is, it appeared that a new lineage had sprung out of the archaebacteria lineage and had slowly developed such unique characteristics as a nucleus and the various other compartments in our cells, along with a range of other features.

Was this really how it happened? Woese and Fox based their analysis on one single gene. What if they had started with another widely distributed gene, such as the gene that allows you to metabolize alcohol? Our human version of that gene corresponds more closely to the eubacterial gene than to the archaebacterial one, the opposite of what we found for the *16S RNA* gene. A tree of life based on the gene for breaking down alcohol would show that humans and other eukaryotes evolved from eubacteria, not from archaebacteria.

Which of these trees is right? There is good evidence that both trees are correct, making for an extremely interesting situation. Each tree accurately describes the evolutionary history of the gene on which it is based, but neither tree is a representation of the evolutionary history of the society of genes of eukaryotes. In Chapter 8, when we discussed horizontal gene transfer among bacteria, we found that individual genes do not necessarily represent the evolutionary history of the whole society. Instead, they may be late immigrants to the society of genes and bring with them their own history. Disagreement among individual gene trees may simply reflect the fact that individual genes have spent parts of their history in different societies of genes.

Then, something even more surprising was discovered.

If You Can't Beat Them, Join Them

To understand why the tree of life drawn up by Woese and Fox would have been radically different if they had used the alcohol-metabolizing gene instead of *16S RNA*, we have to take a step back. Cells as large as our own come with big costs. A cell is a crowded place, teeming with molecular machines, their accessories, their raw materials, and their products. The bigger the cell, the more it contains. A cell needs energy to run its business, and cells of different sizes have energy requirements proportional to their volumes.

Where does that energy come from? Bacterial cells are surrounded by a wall made up of long sugar molecules crosslinked with proteins. That outer wall and the cell's interior are separated by a membrane. To generate the cellular energy currency—an energy-packed molecule called ATP—bacteria burn sugars or capture sunlight, thereby pumping protons (the nuclei of hydrogen atoms) from the inside of the cell into the space between the cell's outer wall and the membrane. The positively charged protons accumulate there, and the mutual repulsion, caused by their electrical charge, drives them back into the cell. The space between the membrane and the cell wall fulfills the same function as a pond feeding a mill. The proton backflow is harnessed by special proteins in the membrane that act like a mill's waterwheels, charging ATP molecules with energy obtained from the proton flow.

Because the only membranes that eubacteria and archaebacteria possess are their cellular membranes, the maximum energy a bacterium can produce is proportional to its surface area. This energy is sufficient to run a cell the size of that bacteria. Now here comes a problem: surfaces don't grow as fast as volumes. A cell that is scaled up by doubling its diameter will have a fourfold increase in surface area, but an eightfold increase in volume. That's easiest to see if we imagine a cell as a cube: in doubling each edge, the two-dimensional faces become 2×2 times bigger, while the three-dimensional volume increases $2 \times 2 \times 2$ times. Energy requirements, which scale with the volume, thus grow much faster than energy production, limited by the cellular surface. Beyond a certain size—one well below the size of one of your cells—eubacteria and archaebacteria cannot support their own energy needs.

How, then, did our ancestors solve this seemingly insurmountable problem? Why is it that you can sustain cells large enough to build a brain? The same principles that restrict the size of bacteria are at work in your own cells: their energy supply has to be sustained by a membrane larger than their surface. This missing surface that lurks in the shadows is the Irving of multicellular life.

The critical difference between bacteria and you is that your cells do not use their outer membranes to supply energy. Instead, they use the surfaces of special structures located within the cell called mitochondria. These are one type of specialized compartment or "workshop" present in all cells of eukaryotes. Each eukaryotic cell has many of these mitochondria, our cellular power plants.

The building and the workings of all the different compartments in your cells are completely specified by the genes on your forty-six chromosomes, with the exception of mitochondria; each mitochondrion possesses its own little genome. The mitochondrial chromosome is rather different in structure from your other chromosomes: it is circular, like the typical chromosomes of bacteria.

How did such a partially independent structure in the cell evolve? In 1970, biologist Lynn Margulis put forth a bold theory. She proposed that mitochondria were once independent eubacteria. At some point, an early eukaryotic cell swallowed a eubacterium, but instead of digesting it, it allowed it to live on, dividing and multiplying inside its cell. The descendants of the host eukaryotic cell and the eubacterium have lived in happy symbiosis ever since (Figure 9.4).

Initially, few were convinced by Margulis's theory, but as more and more genomic sequences accumulated, the evidence in its favor became overwhelming. The most convincing piece

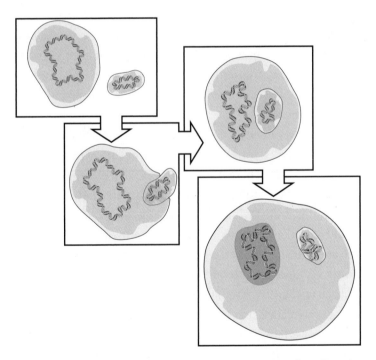

Figure 9.4: In Lynn Margulis's theory, an early eukaryotic cell swallowed a eubacterium. It didn't digest the eubacterium but allowed it to become a tenant, where it lived on and multiplied, evolving into the mitochondrium. Martin and Müller later proposed that the landlord was an archaebacterium, and this union was the origin of eukaryotes.

of evidence was the discovery that the genes inhabiting the mitochondrial genome are most similar to those found in a particular group of eubacteria. In the tree of life, mitochondrial genes do not group with other eukaryotic genes, but with a particular class of eubacteria.

Almost twenty years later, Bill Martin and Miklós Müller proposed a new interpretation of Margulis's theory. At that time, most experts assumed that the cell that acquired the mitochondrion was an early eukaryote, even though no traces of such primitive eukaryotes exist today. Martin and Müller proposed that this ancestor must have been an archaebacterium, and that about 2 billion years ago, this one archaebacterium found an ingenious solution to the energy problem of cell expansion: it engulfed a small eubacterium and turned it into a power source.

The first time the archaebacterium took in the eubacterium but didn't digest it must have been an accident, akin to a major mutation. Maybe these initial stages of what was to become a truly symbiotic relationship were more like that between a parasite and a host. Perhaps it was the small eubacterium who found a way into the comfortable and nutrient-rich interior of the archaebacterium, where it was fed and protected from the harsh and dangerous world outside. Or maybe the two types of bacteria had collaborated in some ways before, now tightening an already existing relationship. Whatever the initial advantages were to each of the two partners, somehow the two survived together and learned how to thrive as a team. The eubacterium would multiply inside the archaebacterium, providing ample energy. The archaebacterium would use this energy, presumably enabling it to colonize environments not accessible to other eubacteria or archaebacteria.

Throughout this book, we have presented the gene as the target of natural selection, but that is actually an oversimplification, intended to help structure our thoughts. Human language and thought have to transform a world of stunning diversity and complexity into less than a million words and a limited set of concepts. To be more accurate, the target of evolution is whatever fulfills the three requirements of natural selection: variation, heritability, and fitness effects. In the origin of mitochondria, the target of selection was the full eubacterium that took up residence inside the archaebacterium.

As soon as one cell swallowed the other, they had two complete genomes between them. Think of the archaebacterium as the landlord and the eubacterium as the tenant. Gradually, the tenant began to lose genes that it no longer required after giving up its autonomous life. These genes suffered mutations from time to time, just as any gene would naturally. But while a debilitating mutation to any gene that contributes to the cell's success would lead to the extinction of its carrier by negative selection, nothing kept these now-defunct genes from degrading, similar to the decay of the scent detector genes we met in Chapter 8. The tenant, who once was a fully independent organism, eventually lost the ability to grow outside its host. Many copies of the tenant filled the cell, and when the host divided, the multiple tenants split between the daughter cells.

Yet the asymmetry between host and tenant only continued. In the cell, if the landlord dies, that spells the end of the tenant, too. But if one of the many tenants dies, the others continue on with business as usual. The dead tenant decays, and what remains of it is eaten up by the cellular machinery.

In some rare cases, parts of the dead tenant's crumbling genome are accidentally pasted into the landlord's genome, just as happens in horizontal gene transfer between different bacteria. This stretch of DNA is then found in both the host's and the tenant's genome, even though one copy was perfectly sufficient to run the mitochondria's operations. The redundancy cannot last long: one of the two copies is bound to mutate by accident and thereby self-destruct. If the landlord's copy acquires a debilitating mutation, then it is as if the accidental integration into its genome never happened. But if the tenant's copy is mutated, a gene previously encoded on the tenant's genome is transferred to the host's genome. In this way, the mitochondrion's genome shrinks over time, irreversibly transferring more and more of its genes to the host.

Each of your mitochondria has its own genome, but, because of this process of attrition, its genome is awfully skimpy. It harbors only thirty-seven genes, while 20,000 genes are tucked inside the nucleus. These thirty-seven genes are not enough to run a machinery as complex as a mitochondrion, which functions almost like a small cell inside your cells. For things to work, each of your mitochondria relies on the collaboration of more than 600 different proteins. The vast majority of the corresponding genes have by now been transferred to one of the forty-six chromosomes in your cells' nucleus, and the proteins they make are imported back into the mitochondrion from the main cell. In some of our very distant eukaryotic cousins, the entire mitochondrial genome has by now been moved to the main genome.

Why do your cells have a nucleus? The nucleus may have arisen as a consequence of the arrival of the mitochondrial tenant. The walls of the nucleus that enclose the host's genome protect it from the constant influx of the deceased tenant's DNA, reducing the constant duplication of mitochondrial

DNA on the central genome. As Robert Frost wrote, good fences make good neighbors.

Before joining forces, the two cells that would become landlord and tenant may have been competitors. Forming a merger benefited both immensely. It enabled their joint descendants to branch out into the amazing forms of multicellular life that we see on the planet today. But you do not need to stop looking for collaborators just because you already have one. If you would benefit from expanding your business in a direction in which you have no previous experience, and if there is another company that knows how to do it, then one way to proceed is to acquire that other company and integrate it into your own organization. This is what the ancestor of all plants and algae did. One early eukaryote engulfed a cyanobacterium that knew how to use the sun's energy to convert carbon dioxide (CO_2) into sugar. To this day, this job is still performed by the resulting second type of tenant, the chloroplasts of plants and algae.

Long Live the Prokaryotes

Some biologists regard lumping eubacteria and archaebacteria into one group—the prokaryotes—to be taboo. Their opinion is that not only are these two groups extremely distant relatives, but also that lumping them together is akin to classifying your brother and cousin as relatives, but not you—it's unnatural. Woese and Fox's tree of life showed that archaebacteria and eukaryotes are more closely related to each other than to the eubacteria (they are "brothers"), so it can be argued that it makes no sense to put the archaebacteria together with the more distantly related eubacteria (their "cousins").

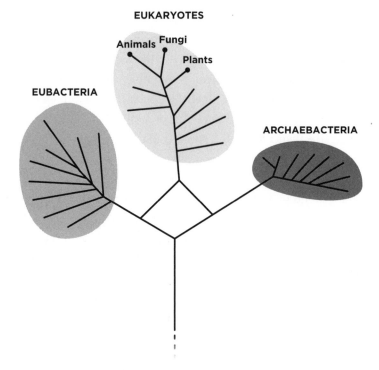

Figure 9.5: About 2 billion years ago, eukaryotes were formed as a merger of a eubacterium and an archaebacterium.

This point of view has a crucial missing piece: the biggest divide in the living world separates the fused cells (the eukaryotes, including you) from the nonfused cells (the prokaryotes). To put it another way: there are normal life forms, the prokaryotes, that happen to come in two flavors—archaebacteria and eubacteria—and then there are weird mixtures of the two, the eukaryotes (Figure 9.5). It just so happens that to evolve into the fabulous forms of animals, plants, and fungi, you need to be a little weird. You need a special kind of collaboration, one in which you team up with your ancient rival. The development of this intimate relationship between an archaebacterium and a eubacterium was the defining step in the evolution of eukaryotes. That is the secret of our success.

Scaling up in sophistication requires collaboration, be it in the cellular world or in politics. When Abraham Lincoln sought the presidential nomination in 1860, he faced three prominent rivals. He captured the prize by becoming everyone's favorite second choice. When time came to fill a cabinet, he chose to surround himself with his former rivals, building a strong cabinet made up of the most qualified people of his day. It may be that this collaboration among a "team of rivals"—as the historian Doris Kearns Goodwin put it—was the secret ingredient for the success of Lincoln's presidency, just as it was at the heart of the success of our eukaryote ancestors.

But the mitochondria are not the only shadowy presence in our society of genes. In Chapter 10, we expose another shadow world that survives at the very core of our society—simply because it can.

Life's Unwinnable War against Freeloaders

Only the dead have seen the end of war.

—Plato

THERE ARE FREELOADERS IN ALL SOCIETIES. REMEMBER KRAMER, Jerry Seinfeld's neighbor? He was constantly freeloading off Jerry. In one episode, when Jerry is accidentally cut and loses a lot of blood, Kramer comes to the rescue by donating blood for Jerry's transfusion. "You've got three pints of Kramer in you, buddy!" Kramer informs Jerry, who has just woken up in the hospital. But Jerry is not altogether pleased: "I can feel his blood inside of me—borrowing things from my blood."

Some of the most ancient and cooperative genes in the human society of genes date back to our eubacterial ancestry, while others are archaebacterial in origin. Every gene has had to earn its keep by contributing to building or running you, the survival machine. That, at least, is the premise we have presented to you so far. But contributing to the community effort is not the only strategy that allows genes to survive. Calculating the number of the DNA letters corresponding to each gene that contributes to the community's success, including the switches needed to manage it all, yields a sum that accounts for less than a third of your genome. A giant proportion of your DNA does not participate in sustaining the society, even if the count includes not only the 20,000 protein-coding genes, which have been the focus of this book, but also other genomic regions that are believed to contribute to your fitness. If the rest does not contribute to human well-being or success, what does this 4-billion-letter-strong majority do, and why has it lasted?

To answer these questions, let's take a closer look at the sequences in this majority. No less than 15 percent of your genome corresponds to one particular sequence of letters, repeated in half a million copies. To put that into perspective, imagine going into the New York Public Library and finding that out of its 12 million books, 1.8 million are virtually identical. What a waste of shelf space! The 500,000 copies in your genome are not perfectly identical: many pairs share more than 99 percent of their letters, while others show somewhat larger differences.

Previously, in discussing similarity of DNA letter sequences, we interpreted them as a sign of shared ancestry, in the way that a person's hair color and nose shape might indicate lineage. The same applies here: common ancestry is the simplest explanation for the similarity among these 500,000 copies. All those elements can trace back their history to one template sequence that arrived in the society of genes of one of our primate ancestors many million years ago (Figure 10.1). Therefore, the copies in your genome comprise one large family, divided into subfamilies, with each individual arising from the duplication of a previous copy. This is very similar to the gene duplications we discussed earlier, although the precise mechanism of duplication differs. The variations among individual members arise from accumulating mutations: each

Figure 10.1: A schematic of a family tree of LINE1 elements. The "leaves" are LINE1 copies that currently exist in the genome. The darker circles show those LINE1 copies that are still functional. Each branching point marks one duplication. The stem of the tree corresponds to the first LINE1 sequence that joined the human society of genes.

copy may acquire slight changes that are then inherited by later, additional copies.

The Bottom Line

These sequences are called LINE1s (*long interspersed elements type 1*). They are genes, but peculiar ones. Each full LINE1 is 6,000 letters long. There are also many shortened copies, with only their ends preserved. The LINE1 sequence of letters is not random; a full LINE1 sequence controls three simple functions that together implement an efficient program. That program encompasses the following jobs: management, RNA-to-DNA converter, and DNA breaker. The management part of the program does not encode a protein; it is a region that mimics a signal that is also found in front of respectable human genes, those that contribute to the society's success. The signal induces the DNA reading machine, the polymerase, to copy the LINE1 into a messenger RNA. Each of the other two functional LINE1 regions encodes a protein. The RNA-to-DNA converter is like a polymerase running backward: it makes a DNA copy from a stretch of RNA. There is also a machine that does that among the respectable society members: the telomerase we met in Chapter 1, which rebuilds the chromosome ends that get shortened in cell division. The other protein encoded by LINE1s, the DNA breaker, is capable of cutting through the double helix of a chromosome. This type of protein is also found among respectable society members; it enables, for example, the recombination required to mix our genome halves in the preparation of sperm and egg cells (Chapter 3).

The program that this three-gene sequence executes goes like this. The promoter recruits the cell's polymerase, which produces an RNA copy of the complete LINE1 sequence. This RNA copy poses as a respectable RNA transcript, and the naïve cellular machinery uses it as a template for making the two proteins. The freshly made RNA-to-DNA converter protein then seizes the very RNA molecule that served to produce it and copies it back into DNA. To identify the LINE1 RNA among the millions of other RNAs in the cell, this protein uses a type of barcode, a specific sequence of letters at the end of the LINE1. Finally, the DNA breaker protein cuts the genome at a random place, where the freshly made DNA copy is then inserted.

In effect, the LINE1 element has copied and pasted itself to another region of the genome, employing the unsuspecting help of some respectable genes—LINEs could have aptly been called "self-copy/pasters" (Figure 10.2, left). By executing their simple program, LINE1s can multiply within a society of genes until their copies outnumber the remaining society members, explaining the hundreds of thousands of copies you find in your genome today.

While the LINE1s do not contribute to the success of the society of genes, they carry out a function—the function of securing their own survival. If a society contained only one LINE1, and this LINE1 were inactivated through a mutation, then that would spell the end of it. If, however, many active LINE1 copies live in the same society of genes, then they are unlikely to be weeded out by accidental mutations. For every LINE1 debilitated by a mutation, several new LINE1s are created by the copy/paste mechanism. As long as LINEs duplicates at a faster rate than the rate of that elimination, the LINE1 family lives to fight another day. Judging from the 500,000 copies present in your genome, LINE1 duplication seems to have run very fast indeed.

Why does the society tolerate such rampant copying? The LINE1 freeloading imposes a burden on the society of genes not only by redirecting the attention of the cellular machinery for DNA reading and protein production towards LINE1's

LINEs

SINEs

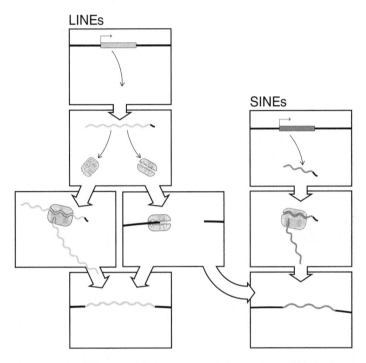

Figure 10.2: LINE1 elements (left) can copy and paste themselves back into the genome, ensuring their proliferation within it. SINE elements (right) manage to have themselves copied and pasted back into the genome by freeloading on the LINE1 proteins.

selfish goals, but also by taking up genomic real estate that needs to be maintained and copied at each cell division. In our case and that of our ancestors, this burden was obviously not large enough for the rest of the society to break down, and so the LINE1s persist. If that burden had been substantially larger, then those individuals carrying many LINE1s in their genome would have left fewer offspring, and the numbers of LINE1s would have dwindled.

All genes jump through the generations with no sentimentality for the individual organism they build. But the success of normal genes hinges on their cooperation. As we have seen, the only way genes can build whole organisms to propel themselves into the next generation is through cooperation; no gene can make it alone. Capitalist societies operate in a similar way: by being selfish within the confines of fair rules, each individual helps maximize the common good. By making a contribution, a normal gene can secure its survival in the society of genes, because if it is lost, the society as a whole suffers.

While all genes may share the selfish motive of making it into the next generation, the LINE1s are a special case: they are freeloaders, pure and simple. They ensure their survival not by making themselves useful but by multiplying in the genome faster than they can be removed. Since this strategy is successful, they do not need to justify their existence in any other way. With their persistence guaranteed, they have no need to contribute to the individual in whom they are housed.

Remember that only 30 percent of your genome is occupied by respectable genes. Of the 70 percent that remain, the LINE1s account for less than one quarter. What's going on in the rest of your genome? Part of it is taken up by another family of freeloaders, the Alu. Your genome is home to a full 1 million copies from that family. Each Alu is between 100 and 400 DNA letters long. More numerous but shorter than LINE1s, Alus amount to 10 percent of your genome. How do Alus ensure their survival? They belong to an extended family called SINEs (short interspersed elements). Alus are similar to LINE1s, but are missing part of the LINE1 sequence. This difference hints at their strategy: though the two are closely related, Alus are even more insidious than LINE1s.

Like the LINE1s, each Alu has at its beginning a "read me" signal that fools the polymerase into making an RNA copy. In addition, each Alu has exactly the same short barcode sequence as LINE1s. That is where the similarity between them ends. In fact, these two signals are all that an Alu has. It does not encode any proteins, and it has no RNA-to-DNA converter or DNA breaker. How does an Alu self-propagate without them?

Recall that the LINE1 RNA-to-DNA converter uses a barcode to identify its own RNA, and that all Alus have an identical copy of that barcode. This fools the LINE1 protein machinery into mistaking an Alu RNA for a LINE1 copy, converting it into DNA, so that it can be inserted into a cut made by the LINE1 DNA breaker. Thus, the Alus are not only freeloading off the respectable society members—on top of that, they are also freeloading off the LINE1s! (Figure 10.2, right). The success of this strategy is evidenced in the huge number of copies that Alus have generated by taking free-loading to another level.

How did Alus evolve? One possibility is that one LINE1 element accidentally lost its protein-coding part, but managed to live on through the help of its intact relatives. There is a second possible scenario. Imagine a duplicate copy of a re-spectable gene—a redundant society member whose fate would be of no consequence for the society. Now assume that a LINE1 element was by chance only partially copied back into DNA, and the end of this LINE1 element containing the barcode was inserted into the redundant copy of the respect-able gene. The gene already had a signal to attract the cellular machinery for reading DNA. Accidentally, this new gene became a master of freeloading: it was read by the cell's polymerase, copied into DNA, and inserted by the LINE1 proteins. It indeed looks like Alus arose through the latter scenario, because the Alu's "read me" signals are similar to those of other genes.

Alus are not unique in freeloading off other freeloaders. If one society member figured out a way to beat the system, why should others not follow suit? Just as Alus piggyback on the LINE1s, another family of freeloaders, called MIR, scrounge their copies from another family of LINEs (LINE2s). All families of SINEs and LINEs together occupy a full third of your genomic real estate, and they outnumber your respectable genes by 10,000 fold. And your genome's burden of freeloading doesn't stop there. There are many other families of freeloaders, each exploiting the society of genes by hitching a ride through the generations without contributing to the society.

While about two-thirds of our genomic real estate is occupied by freeloaders, we can still consider ourselves lucky in comparison to other species. The modest onion has a genome blown up to 30 billion letters, five times as large as our own, and some amoeba have genomes one hundred times larger than those of humans. Most of those ridiculously numerous DNA letters are parts of freeloading genes, just as the LINEs and SINEs in your own genome. The tolerated amount of freeloaders depends on the organism's lifestyle. So what do we have in common with onions and amoebae? The accumulated junk in our genomes simply shows that we all lived in small populations for most of our evolutionary history. In small populations, chance plays a bigger role, and natural selection is less efficient. Thus, selfish society members that put a small load onto their carrier's back have a better chance to survive. To some extent, the size of our genome reflects the fact that humans used to live in small groups until a few thousand years ago.

The LINE1 DNA breaker cuts your chromosomes at more or less random places, yet the LINE1 and Alu repeats are not equally distributed across your whole genome. For example, the regions in the genome that contain the large HOX gene clusters we met in Chapter 7, responsible for setting up our body plan in embryonic development, are almost completely devoid of freeloaders. Are the freeloaders intentionally respectful of those regions, because they somehow know that disrupting them would destroy the organism they rely on for their survival? Unlikely. The freeloaders are blind fraudsters that target all regions equally, just like random mutations that cause single letters to change. But as in the case of other mutations, negative selection kicks in. Each of those millions of times that a freeloader copy was inserted into a HOX gene led to a genome incapable of building a working organism. The insertion spelt total disaster for the complete genome, bringing the hopeful new freeloader copy down with it.

The Spandrels of San Marco

But does freeloading DNA really serve no function for the society? Does the shady origin of LINE1s and Alus preclude them from at least sometimes taking up new and productive professions? Given how natural selection will seize on any variation that happens to provide fitness benefits, it would seem unlikely that the integration of millions upon millions of freeloaders into your genome would never lead to anything useful. Indeed, such useful functions continue to be detected for more and more individual freeloaders. Some of them facilitate gene evolution, where a gene is expanded by the insertion of a freeloader. In other instances, a freeloading element may change when a gene is read, by being inserted into a gene's control region and altering the molecular switches used by the gene's managers to turn it on and off. In the rare case that this improves the gene's contribution to the fitness of its carriers, this particular freeloader itself will become a part of a respectable society member. In yet other examples, the "read me" signal of an inserted freeloading element may allow a new gene to attract the polymerase machinery and thereby become useful.

Let us consider an altogether different example: the flippers of penguins. These evolved from wings that were no longer used for flight. The eminent evolutionary biologist Stephen J. Gould would have said that the flippers were exapted—they were modified from an organ that previously evolved for an entirely different purpose. In the same way, a freeloading DNA that accidentally takes up a new function is also a case of exaptation. Its original design followed from a strategy for duplicating itself. After its accidental insertion at the right place in the genome, it was exapted for a purpose that proved to be advantageous to the organism. It is important to realize that originally, the only reason for the element's continued existence was freeloading. Since the vast majority of freeloaders do not exert any advantageous effects, the best explanation for their continued existence is not that they perform any useful roles but simply that they continue to be good at copying themselves. Like Kramer, they are freeloaders, even if sometimes their actions can be beneficial. The continued integration introduces a stream of new variation into the society of genes—and among that huge variety, it is inevitable that occasionally, one freeloader lends itself to something useful.

That the urge to ascribe a function to a freeloader is still manifest among biologists reveals more about psychology than about biology. We may want to believe that our society of genes evolved as an efficient organization, not the junky and disorderly mess we have described in this chapter. Any good scientific paper tells a gripping story, and a common storyline for papers on LINEs or SINEs is this: we thought that all LINEs (or SINEs) are freeloading junk, but instead, we find that this or that copy has a fitness-enhancing function. This clearly makes for a good narrative, but its implication that all junk is functional is misleading.

Stephen J. Gould sought to expose the preconception that everything in biology must have an adaptive advantage. He argued that structures may have features that do not serve any purpose in themselves—rather, these features reflect specific constraints. Gould illustrated this idea with an architectural motif, a spandrel, the triangular section between two arches or between an arch and a rectangular enclosure (Figure 10.3). Churches like that of San Marco in Venice tend to cover these spandrels with lavish decorations. Of course, providing space for those decorations is not the spandrel's original function: it was a consequence of structural constraints and was then exapted for decorative purposes. Similarly, the freeloaders of our

Figure 10.3: A spandrel—the triangular space between two arches—exists for structural reasons, but is frequently exapted for use as a decorative element.

society of genes are a phenomenon that arose by natural selection in the genome. Some of them may have been exapted
for other uses, but that should not confuse us about their origin as pure freeloaders.

Life's Oldest Enemy

SINEs and LINEs just provide us with a glimpse of how the
ubiquitous phenomenon of freeloading works. The way freeloaders take advantage of the society of genes may remind you
of the viruses we met in Chapter 2. Viruses are the ancient
masters of freeloading, forming a vast and surprisingly complex shadow world beyond the cellular life that we are ac-
customed to. This virus world is in fact so enormous that the
overwhelming majority of all genomes on earth are those of
viruses, which, while being small, outnumber the genomes of
cells by ten to one.

Similar to LINEs and SINEs, a virus will redirect your
cell's machinery into making copies of the virus. Sometimes,
though, viruses may instead go into hiding, waiting for the
right time to attack. A herpes virus may sneak into one of your
nerve cells in camouflage, carefully escaping the attention of
your body's immune system. Instead of setting to work right
away, the virus slips into a "sleeper" stage, lying dormant for
months or years. The virus may then wake up, perhaps by
noticing that your immune system has its hands full with an
influenza infection. It then travels to your skin, where it mul-
tiplies and causes cold sores, blisters that may break open to
spurt out viral copies. Upon close contact with anyone, the tiny

viral genome jumps to a new and unsuspecting home, and the evil cycle begins again.

Some types of viral infections are also important risk factors for the development of specific cancers. Most viruses that cause cancer do so either by directly encoding a gene that makes the infected cell override one of the body's eight defenses against cancer (remember Chapter 1), or by inserting a viral sequence in the control region of a human gene that can contribute to cancer (a potential oncogene), changing its expression level. In this way, the viruses cause the infected cells to complete one step toward cancer. Because there are several other steps to cancer that have to be completed, not everybody infected with these viruses evolves a cancer. For example, most carriers of human papilloma viruses, which sometimes cause genital warts, do not experience serious consequences. But practically all incidences of cervical cancers—the second most frequent cancer in women—are initiated by such an infection. When the cancerous cells divide, the daughter cells also inherit viral copies. In this way, specific viruses may benefit from causing cancer by expanding the facilities for their replication.

Viruses do not attack only humans and other animals or bacteria. They are the scourge of all cellular life forms, including plants, fungi, and single-celled organisms. In eubacteria, sleeper viruses sometimes employ an even more sinister strategy than in the case of herpes. After secretively entering a bacterial cell, they insert their own DNA into the bacterial genome. Whenever the bacterium divides, its daughter cells inherit a faithful copy of the virus inside their genomes. As long as life is easy for the bacteria, their happy reproduction benefits the sleeper virus. But when trouble comes along, perhaps in the form of starvation, the sleepers decide that it is time to make an escape before their host dies. They wake up and hijack the cellular machinery, turning it into a factory that churns out new viruses until the bacterium dies of exhaustion or is eventually killed by those same viruses when they make their escape.

Viruses show a spectacular variety that far outstrips even the amazing diversity seen among cellular life. An intriguing case in point is the diversity of data storage systems across virus genomes. All cellular life stores genomic information in the form of the double-stranded DNA helix. In principle, the same information could be stored as single-stranded DNA, as single-stranded RNA, or as double-stranded RNA—but that's not what any animal, plant, fungus, or bacterium uses. All cellular life forms use RNA only transiently as messengers or for specific functions in the cell. In viruses, however, all the different possible genetic storage systems are used. Some viruses use double strands of DNA; others use a single, unpaired, strand; and yet others use double- or single-stranded RNA. Together with different ways to read the genes off single-stranded DNA or RNA, this amounts to a total of seven distinctly different systems.

Viruses come in millions of different shapes and sizes, showing an exuberance unseen in cellular life. Strikingly, there is not even a single gene in common between all viruses. This, again, is in sharp contrast to cellular life. Approximately fifty genes are universal to the brotherhood and sisterhood of cells: the genes that encode the machinery for unpacking DNA, the polymerase for reading DNA, the ribosome for making proteins—all these are members of the societies of genes of all cellular life forms. Why is something similar not true for viruses? Why, for example, is there no universal viral gene that encodes the hull, the protein shell of the virus? Not only do different viruses have different ways to make a hull, but some viruses do without a hull gene at all. These are the freeloaders of freeloaders, parasites not only of human cells but also of their viral cousins. Invading cells together with their cousins, these crooks use their cousin's instructions for the hull protein for their own purposes, just as the Alus freeload off the LINE1s. The lack of coherence among viral genes may be due to their freeloading nature: in principle, they can delegate each and every essential function to the genes of their victims.

Most viral genomes encode maybe a handful of genes to supply those functions not already encoded by their host. However, there are also huge viruses with genomes up to a million letters long, harboring over 1,000 genes. These giant viruses are of similar complexity as some freeloading bacteria that can only live inside the cells of their hosts, and these giant viruses use similar strategies for their propagation as their bacterial counterparts. One crucial difference is that the viruses—giant or not—forsake their own hull to invade the attacked cells with their genomes only, while bacterial parasites move into their unfortunate hosts intact, their genome never leaving their own cell walls.

In principle, however, viral genomes need not even contain protein-encoding genes at all. One example is a viroid. A viroid has a life style very similar to that of viruses, but it does not package itself into a hull—viroids are free-floating genomes, made up of RNA. Their genomes are only about 300 letters long, and they do not specify any proteins at all. Instead, the viroid genome just contains operating instructions that manipulate their victim's machinery into making viroid copies, transmitted directly in the form of RNA. You do not need to fear them, though: viroids have so far only been found to infect plants.

As far as we can tell, all cellular life stems from a common ancestral society of genes that made its living at the dawn of life. A compelling case for this scenario is made by the existence of the about fifty genes universal to all cellular life forms. But what about viruses? Do they trace back to the same ancestral virus? And what came first: the virus or the cell? Since viruses rely on our proteins to "survive," it is hard to imagine a virus-only world. But evidence suggests that cells did not arise first either. Instead, an epic war likely raged from the beginning, from a common origin of cells and their freeloaders.

Biology for Beginners

Life as we know it today is complex, but in the beginning it must have been very simple—otherwise it would have been too unlikely to appear. In today's cells, DNA and RNA store information, while proteins carry out most of the molecular functions inside cells. Which of these types of molecules evolved first? The DNA, which makes information heritable, or the proteins, which process that information? Is this a classic chicken-and-egg question? We do not know the exact answers. There are competing ideas, the jury is still out, and given that we are talking about something that happened 4 billion years ago at an unknown location, we may never be entirely sure.

But to get an idea how life *could* have evolved, let's look at what may be our current best guess.

Because the RNA letters A and U (remember that the latter corresponds to T in DNA), as well as the letters G and C, prefer to bind to each other, the sequence of RNA letters can fold back onto itself. Through these folds, an RNA molecule assumes a three-dimensional shape that is encoded in the sequence of letters, just like the amino acid chain that makes up a protein spontaneously folds into its protein structure. Depending on the exact shape of the folded RNA, the resulting molecule can become a tiny molecular machine, such as an enzyme that promotes specific chemical reactions. So, many functions that are performed by proteins today can in principle also be performed by RNA molecules. A sizeable chunk of the protein production machinery of today's organisms, the ribosome, still is made from RNA.

Thus, RNA can simultaneously encode heritable information and act as a molecular machine. This dual role makes it possible, at least in principle, for one string of RNA to carry the heritable information *and* to catalyze its own reproduction (Figure 10.4). Thus, initially, no clear distinction may have existed between molecules encoding genetic information and molecules performing functions. A "living" world may have consisted entirely of RNA, in a time before DNA, proteins, and cell walls existed.

Where could such RNA replicators have evolved? Life requires a source of energy, and all energy that could possibly be used to sustain life ultimately comes from one of two sources: the rays of the sun (exploited by means of photosynthesis by plants, algae, and some bacteria); or chemical energy produced by geothermal processes. Photosynthesis requires a

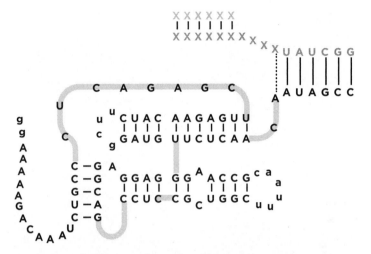

Figure 10.4: An RNA molecule capable of replicating RNA. The parallel horizontal and vertical lines indicate the binding between complementary RNA letters (A-U and C-G), which gives the RNA molecule its characteristic shape. The RNA letters shown in gray represent a stretch of RNA that is being copied, with X standing for arbitrary letters. Redrawn with modifications from Wochner et al. 2011.

complex, specialized machinery, making it an unlikely candidate for the support of the earliest life. But the chemistry that occurs when hydrothermal vents at the bottom of the oceans spew out hot, chemical-rich fluids into cold seawater already resembles important parts of the central metabolism found in today's life forms.

Cellular energy production needs a membrane with different concentrations of protons on either side—remember that this is why plants and animals need mitochondria to support their lavish lifestyles (Chapter 9). But proton concentration differences that naturally form around deep-sea vents may provide a very similar effect. It thus seems plausible that life started in little cavities in the rocks that surround

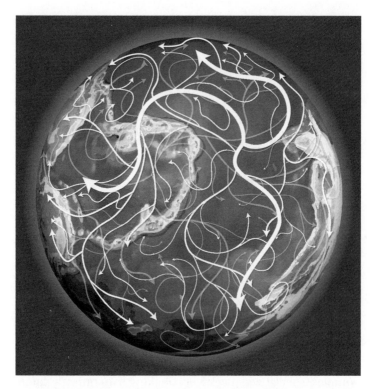

Figure 10.5: Life may have first arisen in small rock cavities at a deep-sea vent, forming a net of early societies of genes that likely consisted of loose collections of RNA molecules. With the invention of cell walls, life was freed from its rock home and started to conquer the oceans and later the land masses of our world, evolving into the myriad shapes we see around us today.

certain hydrothermal vents on the sea floor. We can imagine that the first RNA molecules assembled spontaneously in this rich chemical environment, forming an early, primitive society of genes. Given enough time, the first RNA-replicator may have arisen. Once in existence, it would have started to replicate itself.

In this earliest society of genes, freeloading must have been particularly rampant. The first RNA molecules that could replicate themselves were very likely incapable of clearly distinguishing between their own kind and other RNA sequences. Thus, they would have replicated not only themselves but also other RNA that they came across. These RNA freeloaders would have multiplied alongside the replicators, putting a massive additional burden on the replicators. These freeloaders would have been the first viroids, forerunners of the later viruses. To relieve themselves from this burden, the replicators had to keep the freeloaders away from themselves. The first replicator that managed to surround itself with a membrane while still letting food in had a massive advantage over its peers still burdened by the freeloaders. The cell wall thus invented had an interesting side effect: it made its carriers independent of the rock cavities that kept them together. It formed a vessel, allowing the society of genes inside to venture out into the surrounding ocean and beyond (Figure 10.5). The rest is natural history.

Epilogue

As this whole volume is one long argument, it may be
convenient to the reader to have the leading facts
and inferences briefly recapitulated.

—Charles Darwin

DARWIN REFERRED TO HIS MAGNUM OPUS, *ON THE ORIGIN OF*
Species, as "one long argument." He knew that his claim that
all life evolved from a common source by natural selection was
an extraordinary one and required unimpeachable evidence
for it to be taken seriously. He laid out the principle of natural
selection at the beginning of the book, and then proceeded to
support it with examples from geology, fossils, animal breeding,
developmental biology, and taxonomy. He carefully arranged
these examples to construct a picture of evolution of overwhelming clarity. It was his elaborately arranged argument
that earned Darwin the bulk of credit for discovering natural
selection, even though others—most notably, his contemporary
Alfred Russel Wallace—had independently come up with similar concepts.

In the tradition of a long argument, this book exhibits
the explanatory power that comes from viewing the genetic
makeup of a species as a society of genes. We see the gene as
the target of natural selection, as did Dawkins in his "selfish
gene" theory. However, our approach shifts the focus to the

relationships among the genes as they cooperate and compete in running us, their survival machines. Our genes form alliances as they encode such processes as meiosis or our defense systems. They build a tangled web of associations, where each one can lend a hand to multiple processes. The society of genes cannot stand still, even if it often evolves only by the pervasive powers of chance. This constant change results in the formation of a new society—the splitting off of a new species—whenever a part of the society is separated from it for a sufficient amount of time. But on rare occasions, societies can fuse, jumping up to new orders of complexity. Change also occurs when new members are introduced to the society through duplications or by immigrants from other societies. Of the many successful interactional strategies a gene can employ to stay in the business of life, one of the most proliferative is freeloading.

Throughout the book, we focused on how interactions in the society influence each gene's success, that is, the "economical" view of the society of genes. But at the same time, we also emphasized the historical perspective. Living organisms are the product of their evolutionary history, or, as physicist-turned-biologist Max Delbrück put it: "Any one cell represents more an historical than a physical event . . . any living cell carries with it the experiences of a billion years of experimentation by its ancestors." Our long argument has spanned many levels of history. Starting with the contemporary history of evolution in action inside an organism (Chapters 1 and 2), we moved on to family history (Chapter 3), the "national" history of populations (Chapters 4 and 5), the making of new species (Chapters 6 and 7), the making of animals (Chapter 8), the historical turning point marked by the first eukaryotic cell

(Chapter 9), and, finally, the beginning of life (Chapter 10). In this travel backward, we have shown how viewing genes as a society is a useful framework at all evolutionary timescales. Throughout our journey through life's history, bacteria accompanied us as a point of reference.

What can we learn from that history? On one level, we are a product of our society of genes. So many of our physical attributes, including the structure of the brain, are the product of our alleles. In that light, how should we, as conscious beings, think of ourselves? Our genes influence our thoughts, our feelings, and our impulses. As we saw, a single allele may in principle be enough to bias any organism against other populations of its own kind. It is healthy to realize that our own biases are evoked to aid our selfish alleles, not necessarily to serve us as conscious individuals or humanity as a whole. As Bill Clinton sensed, the tiny differences among human individuals are eclipsed by what we have in common.

We see ourselves as more than soulless survival machines enslaved by their genes. Wherever our judgment is biased by our genes, we must decide whether we want to play along or take a stand. The scarcity of resources throughout most of the society of genes' history has selected for alleles that nudge us toward maximizing our own resources at the cost of others, and this may, for example, contribute to an impulse to simply pass by a homeless person on the street. But instead of blindly playing along with such impulses, we can consciously decide whether we think it more humane to instead offer a greeting or even some help.

The society of genes influences our judgments and choices far beyond the few fundamental processes discussed in this book. Biases in decision making, such as the halo effect that leads us to make overconfident extrapolations from extremely limited information, are also inscribed in our genes. As Daniel Kahneman argued in his book *Thinking, Fast and Slow*, we can vastly improve our decision making if we are aware of such biases and adjust our thought process accordingly. To use our full potential as conscious individuals, we need to recognize not only the biases examined in decision theory, but also all other biases that have evolved in the million-year-old history of our society of genes. There are arenas in which we do well by playing along with our biases, such as our genetically determined disgust with certain toxic smells. Then there are times to choose to consciously act against the biases caused by our genes, as when we campaign against racism.

We live in interesting times. For millions of years, our ancestors have played along with the society of genes, and all other life on Earth apparently still does. But we have started to transcend part of our heritage by slowly widening the circle of what we consider worthy of our protection—from the family to villages and nations, on to humanity as a whole, and beyond, when we think about animal rights.

To paraphrase an old hymn, it is the society of genes that brought us thus far, but it is our humanity that must now bring us home.

Further Reading

1. Evolving Cancer in Eight Easy Steps

Buffenstein, R. 2008. Negligible senescence in the longest living rodent, the naked mole-rat: Insights from a successfully aging species. *Journal of Comparative Physiology B* 178:439–445.

Coyne, J. A. 2009. *Why evolution is true.* New York: Viking.

Darwin, C. 1897. *The origin of species by means of natural selection, or the preservation of favoured races in the struggle for life.* London: J. Murray.

Dawkins, R. 1996. *The blind watchmaker: Why the evidence of evolution reveals a universe without design.* New York: Norton.

Dennett, D. C. 1995. *Darwin's dangerous idea: Evolution and the meanings of life.* New York: Simon & Schuster.

Hanahan, D., and R. A. Weinberg. 2011. Hallmarks of cancer: The next generation. *Cell* 144:646–674.

Krebs, J. E., B. Lewin, S. T. Kilpatrick, and E. S. Goldstein. 2014. *Lewin's genes XI.* Burlington, MA: Jones & Bartlett Learning.

Lander, E. S., L. M. Linton, B. Birren, C. Nusbaum, M. C. Zody, J. Baldwin, K. Devon, K. Dewar, M. Doyle, W. FitzHugh, et al. 2001. Initial sequencing and analysis of the human genome. *Nature* 409: 860–921.

Lynch, M. 2007. *The origins of genome architecture.* Sunderland, MA: Sinauer Associates.

Tabin, C. J., S. M. Bradley, C. I. Bargmann, R. A. Weinberg, A. G. Papageorge, E. M. Scolnick, R. Dhar, D. R. Lowy, and E. H. Chang. 1982. Mechanism of activation of a human oncogene. *Nature* 300:143–149.

Venter, J. C., M. D. Adams, E. W. Myers, P. W. Li, R. J. Mural, G. G. Sutton, H. O. Smith, M. Yandell, C. A. Evans, R. A. Holt, et al. 2001. The sequence of the human genome. *Science* 291:1304–1351.

Watson, J. D. 2008. *Molecular biology of the gene.* San Francisco: Pearson.

Weinberg, R. A. 1998. *One renegade cell: How cancer begins.* New York: Basic Books.

———. 2007. *The biology of cancer.* New York: Garland Science.

Wolchok, J. D. 2014. New drugs free the immune system to fight cancer. *Scientific American* 310, no. 5. http://www.scientificamerican.com/article /new-drugs-free-the-immune-system-to-fight-cancer/.

2. How Your Enemies Define You

Barrangou, R., C. Fremaux, H. Deveau, M. Richards, P. Boyaval, S. Moineau, D. A. Romero, and P. Horvath. 2007. CRISPR provides acquired resistance against viruses in prokaryotes. *Science* 315:1709–1712.

Bartick, M., and A. Reinhold. 2010. The burden of suboptimal breast-feeding in the United States: A pediatric cost analysis. *Pediatrics* 125: e1048–1056.

Freeland, S. J., R. D. Knight, L. F. Landweber, and L. D. Hurst. 2000. Early fixation of an optimal genetic code. *Molecular Biology and Evolution* 17:511–518.

Goldsby, R. A., T. K. Kindt, B.A. Osborne, and J. Kuby. 2003. *Immunology.* 5th ed. New York: W. H. Freeman and Company.

Iranzo, J., A. E. Lobkovsky, Y. I. Wolf, and E. V. Koonin. 2013. Evolutionary dynamics of the prokaryotic adaptive immunity system CRISPR-Cas in an explicit ecological context. *Journal of Bacteriology* 195:3834–3844.

Janeway, C. A., P. Travers, M. Walport, and M. Shlomchik. 2001. *Immuno-biology.* 6th ed. New York: Garland Publishing.

Jones, S. 2000. *Darwin's ghost: The origin of species updated.* New York: Random House.

Judson, H. F. 1996. *The eighth day of creation: Makers of the revolution in biology.* Plainview, NY: CSHL Press.

Levy, A., M. G. Goren, I. Yosef, O. Auster, M. Manor, G. Amitai, R. Edgar, U. Qimron, and R. Sorek. 2015. CRISPR adaptation biases explain preference for acquisition of foreign DNA. *Nature* 520:505–510.

Makarova, K. S., Y. I. Wolf, and E. V. Koonin. 2013. Comparative genomics of defense systems in archaea and bacteria. *Nucleic Acids Research* 41: 4360–4377.

Mezrich, B. 2004. *Bringing down the house: How six students took Vegas for millons.* London: Arrow.

Rechavi, O., L. Houri-Ze'evi, S. Anava, W. S. Goh, S. Y. Kerk, G. J. Hannon, and O. Hobert. 2014. Starvation-induced transgenerational inheritance of small RNAs in C. elegans. *Cell* 158:277–287.

Sander, J. D., and J. K. Joung. 2014. CRISPR-CAS systems for editing, regulating and targeting genomes. *Nature Biotechnology* 32:347–355.

Sorek, R., V. Kunin, and P. Hugenholtz. 2008. CRISPR—A widespread system that provides acquired resistance against phages in bacteria and archaea. *Nature Reviews Microbiology* 6:181–186.

Stern, A., L. Keren, O. Wurtzel, G. Amitai, and R. Sorek. 2010. Self-targeting by CRISPR: Gene regulation or autoimmunity? *Trends in Genetics* 26:335–340.

World Health Organization (WHO). *Breastfeeding* 2015. http://www.who .int/topics/breastfeeding/.

3. What's the Point of Having Sex?

Baym, M., T. Lieberman, E. Kelsic, R. Chait, and R. Kishony. 2015. The bacterial evolution experiment was carried out by these scientists at Harvard medical school.

Burt, A., and R. Trivers. 2006. *Genes in conflict: The biology of selfish genetic elements.* Cambridge, MA: Belknap Press of Harvard University Press.

Dawkins, R. 1976. *The selfish gene.* Oxford: Oxford University Press.

Diamond, J. M. 1997. *Why is sex fun? The evolution of human sexuality.* New York: HarperCollins.

Ellegren, H. 2011. Sex-chromosome evolution: Recent progress and the influence of male and female heterogamety. *Nature Reviews Genetics* 12:157–166.

Flot, J. F., B. Hespeels, X. Li, B. Noel, I. Arkhipova, E. G. J. Danchin, A. Hejnol, B. Henrissat, R. Koszul, J. M. Aury, et al. 2013. Genomic evidence for ameiotic evolution in the bdelloid rotifer *Adineta vaga*. *Nature* 500:453–457.

Haber, J. E. 2013. *Genome stability: DNA repair and recombination*. New York: Garland Science.

Holman, L., and H. Kokko. 2014. The evolution of genomic imprinting: Costs, benefits and long-term consequences. *Biological Reviews* 89:568–587.

Kuroiwa, A., S. Handa, C. Nishiyama, E. Chiba, F. Yamada, S. Abe, and Y. Matsuda. 2011. Additional copies of *CBX2* in the genomes of males of mammals lacking *SRY*, the Amami spiny rat *(Tokudaia osimensis)* and the Tokunoshima spiny rat *(Tokudaia tokunoshimensis)*. *Chromosome Research* 19:635–644.

Murdoch, J. L., B. A. Walker, and V. A. McKusick. 1972. Parental age effects on the occurrence of new mutations for the Marfan syndrome. *Annals of Human Genetics* 35:331–336.

Ridley, M. 2003. *Nature via nurture: Genes, experience, and what makes us human*. New York: HarperCollins.

———. 2011. *Genome: The autobiography of a species in 23 chapters*. New York: MJF Books.

Stearns, S. C. 2009. *Principles of evolution, ecology and behavior*. http://oyc.yale.edu/ecology-and-evolutionary-biology/eeb-122.

United Nations Population Fund. 2011. *Report of the international workshop on skewed sex ratios at birth: Addressing the issue and the way forward*. New York: UNFPA.

Zimmer, C. 2008. *Microcosm: E. Coli and the new science of life*. New York: Pantheon Books.

4. The Clinton Paradox

Bhattacharya, T., J. Stanton, E. Y. Kim, K. J. Kunstman, J. P. Phair, L. P. Jacobson, and S. M. Wolinsky. 2009. *CCL3L1* and HIV/AIDS susceptibility. *Nature Medicine* 15:1112–1115.

Bollongino, R., J. Burger, A. Powell, M. Mashkour, J. D. Vigne, and M. G. Thomas. 2012. Modern taurine cattle descended from small number of near-eastern founders. *Molecular Biology and Evolution* 29:2101–2104.

Burger, J., M. Kirchner, B. Bramanti, W. Haak, and M. G. Thomas. 2007. Absence of the lactase-persistence-associated allele in early Neolithic Europeans. *Proceedings of the National Academy of Sciences of the USA* 104:3736–3741.

Falush, D., T. Wirth, B. Linz, J. K. Pritchard, M. Stephens, M. Kidd, M. J. Blaser, D. Y. Graham, S. Vacher, G. I. Perez-Perez, et al. 2003. Traces of human migrations in helicobacter pylori populations. *Science* 299:1582–1585.

Ferreira, A., I. Marguti, I. Bechmann, V. Jeney, A. Chora, N. R. Palha, S. Rebelo, A. Henri, Y. Beuzard, and M. P. Soares. 2011. Sickle hemoglobin confers tolerance to *Plasmodium* infection. *Cell* 145:398–409.

Freedman, B. I., and T. C. Register. 2012. Effect of race and genetics on vitamin D metabolism, bone and vascular health. *Nature Reviews Nephrology* 8:459–466.

Graur, D., and W.-H. Li. 2000. *Fundamentals of molecular evolution.* Sunderland, MA: Sinauer Associates.

Hancock, A. M., D. B. Witonsky, G. Alkorta-Aranburu, C. M. Beall, A. Gebremedhin, R. Sukernik, G. Utermann, J. K. Pritchard, G. Coop, and A. Di Rienzo. 2011. Adaptations to climate-mediated selective pressures in humans. *PLoS Genetics* 7:e1001375.

Jablonski, N. G. 2012a. Human skin pigmentation as an example of adaptive evolution. *Proceedings of the American Philosophical Society* 156:45–57.

———. 2012b. *Living color: The biological and social meaning of skin color.* Berkeley: University of California Press.

Lander, E. S. 2011. Initial impact of the sequencing of the human genome. *Nature* 470:187–197.

Levy, S., G. Sutton, P. C. Ng, L. Feuk, A. L. Halpern, B. P. Walenz, N. Axelrod, J. Huang, E. F. Kirkness, G. Denisov, et al. 2007. The diploid genome sequence of an individual human. *PLoS Biology* 5:e254.

Monod, J. 1971. *Chance and necessity; an essay on the natural philosophy of modern biology.* 1st American ed. New York: Knopf.

Weber, N., S. P. Carter, S. R. Dall, R. J. Delahay, J. L. McDonald, S. Bearhop, and R. A. McDonald. 2013. Badger social networks correlate with tuberculosis infection. *Current Biology* 23:R915–916.

Wells, S. 2002. *The journey of man: A genetic odyssey.* New York: Random House.

West, S. A., and A. Gardner. 2010. Altruism, spite, and greenbeards. *Science* 327:1341–1344.

5. Promiscuous Genes in a Complex Society

Bencharit, S., C. L. Morton, Y. Xue, P. M. Potter, and M. R. Redinbo. 2003. Structural basis of heroin and cocaine metabolism by a promiscuous human drug-processing enzyme. *Nature Structural Biology* 10:349–356.

Benko, S., J. A. Fantes, J. Amiel, D. J. Kleinjan, S. Thomas, J. Ramsay, N. Jamshidi, A. Essafi, S. Heaney, C. T. Gordon, et al. 2009. Highly conserved non-coding elements on either side of SOX9 associated with Pierre Robin sequence. *Nature Genetics* 41:359–364.

Collins, F. S. 2010. *The language of life: DNA and the revolution in personalized medicine.* New York: Harper.

Danchin, A. 2002. *The Delphic boat: What genomes tell us.* Cambridge, MA: Harvard University Press.

Franke, A., D. P. McGovern, J. C. Barrett, K. Wang, G. L. Radford-Smith, T. Ahmad, C. W. Lees, T. Balschun, J. Lee, R. Roberts, et al. 2010. Genome-wide meta-analysis increases to 71 the number of confirmed Crohn's disease susceptibility loci. *Nature Genetics* 42:1118–1125.

Ginsburg, G. S., and H. F. Willard. 2013. *Genomic and personalized medicine.* Waltham, MA: Academic Press.

Orel, V. 1984. *Mendel.* New York: Oxford University Press.

Orth, J. D., T. M. Conrad, J. Na, J. A. Lerman, H. Nam, A. M. Feist, and B. O. Palsson. 2011. A comprehensive genome-scale reconstruction of *Escherichia coli* metabolism—2011. *Molecular Systems Biology* 7:535.

Rees, J. L., and R. M. Harding. 2012. Understanding the evolution of human pigmentation: Recent contributions from population genetics. *Journal of Investigative Dermatology* 132:846–853.

Szappanos, B., K. Kovacs, B. Szamecz, F. Honti, M. Costanzo, A. Barysh-nikova, G. Gelius-Dietrich, M. J. Lercher, M. Jelasity, C. L. Myers, et al. 2011. An integrated approach to characterize genetic interaction networks in yeast metabolism. *Nature Genetics* 43:656–662.

Trinh, J., and M. Farrer. 2013. Advances in the genetics of Parkinson disease. *Nature Reviews Neurology* 9:445–454.

Visscher, P. M., M. A. Brown, M. I. McCarthy, and J. Yang. 2012. Five years of GWAS discovery. *American Journal of Human Genetics* 90:7–24.

Weinreich, D. M., N. F. Delaney, M. A. DePristo, and D. L. Hartl. 2006. Darwinian evolution can follow only very few mutational paths to fitter proteins. *Science* 312:111–114.

Yanai, I., and C. DeLisi. 2002. The society of genes: Networks of functional links between genes from comparative genomics. *Genome Biology* 3:research0064.

Zimmer, C. 2008. *Microcosm: E. coli and the new science of life*. New York: Pantheon Books.

6. The Chuman Show

Abi-Rached, L., M. J. Jobin, S. Kulkarni, A. McWhinnie, K. Dalva, L. Gragert, F. Babrzadeh, B. Gharizadeh, M. Luo, F. A. Plummer, et al. 2011. The shaping of modern human immune systems by multiregional admixture with archaic humans. *Science* 334:89–94.

Barton, N. H., D. E. G. Briggs, J. A. Eisen, D. B. Goldstein, and N. H. Patel. 2007. *Evolution*. Cold Spring Harbor, NY: Cold Spring Harbor Laboratory Press.

de Waal, F. B. M. 2001. *Tree of origin: What primate behavior can tell us about human social evolution*. Cambridge, MA: Harvard University Press.

Ely, J. J., M. Leland, M. Martino, W. Swett, and C. M. Moore. 1998. Technical note: Chromosomal and mtDNA analysis of Oliver. *American Journal of Physical Anthropology* 105:395–403.

Green, R. E., J. Krause, A. W. Briggs, T. Maricic, U. Stenzel, M. Kircher, N. Patterson, H. Li, W. Zhai, M. H. Fritz, et al. 2010. A draft sequence of the Neandertal genome. *Science* 328:710–722.

Lalueza-Fox, C., and M. T. Gilbert. 2011. Paleogenomics of archaic hominins. *Current Biology* 21:R1002–1009.

Lynch, M. 2010. Rate, molecular spectrum, and consequences of human mutation. *Proceedings of the National Academy of Sciences of the USA* 107:961–968.

Mikkelsen, T. S., L. W. Hillier, E. E. Eichler, M. C. Zody, D. B. Jaffe, S. P. Yang, W. Enard, I. Hellmann, K. Lindblad-Toh, T. K. Altheide, et al. 2005. Initial sequence of the chimpanzee genome and comparison with the human genome. *Nature* 437:69–87.

Pääbo, S. 2015. *Neanderthal man: In search of lost genomes*. New York: Basic Books.

Patterson, N., D. J. Richter, S. Gnerre, E. S. Lander, and D. Reich. 2006. Genetic evidence for complex speciation of humans and chimpanzees. *Nature* 441:1103–1108.

Reich, D., R. E. Green, M. Kircher, J. Krause, N. Patterson, E. Y. Durand, B. Viola, A. W. Briggs, U. Stenzel, P. L. Johnson, et al. 2010. Genetic history of an archaic hominin group from Denisova Cave in Siberia. *Nature* 468:1053–1060.

Specter, M. 2012. Germs are us. *The New Yorker*, October 22.

7. It's in the Way That You Use It

Bateson, W. 1894. *Materials for the study of variation treated with especial regard to discontinuity in the origin of species*. New York: Macmillan.

Benko, S., C. T. Gordon, D. Mallet, R. Sreenivasan, C. Thauvin-Robinet, A. Brendehaug, S. Thomas, O. Bruland, M. David, M. Nicolino, et al. 2011. Disruption of a long distance regulatory region upstream of sox9 in isolated disorders of sex development. *Journal of Medical Genetics* 48: 825–830.

Carroll, S. B. 2005. Evolution at two levels: On genes and form. *PLoS Biology* 3:1159–1166.

Enard, W., P. Khaitovich, J. Klose, S. Zollner, F. Heissig, P. Giavalisco, K. Nieselt-Struwe, E. Muchmore, A. Varki, R. Ravid, et al. 2002. Intra- and interspecific variation in primate gene expression patterns. *Science* 296: 340–343.

Gerhart, J., and M. Kirschner. 1997. *Cells, embryos, and evolution: Toward a cellular and developmental understanding of phenotypic variation and evolutionary adaptability*. Malden, MA: Blackwell Science.

Haesler, S., K. Wada, A. Nshdejan, E. E. Morrisey, T. Lints, E. D. Jarvis, and C. Scharff. 2004. *FoxP2* expression in avian vocal learners and non-learners. *Journal of Neuroscience* 24:3164–3175.

Hunter, C. P., and C. Kenyon. 1995. Specification of anteroposterior cell fates in *Caenorhabditis elegans* by *Drosophila* Hox proteins. *Nature* 377: 229–232.

King, M. C., and A. C. Wilson. 1975. Evolution at two levels in humans and chimpanzees. *Science* 188:107–116.

McLean, C. Y., P. L. Reno, A. A. Pollen, A. I. Bassan, T. D. Capellini, C. Guenther, V. B. Indjeian, X. Lim, D. B. Menke, B. T. Schaar, et al. 2011. Human-specific loss of regulatory DNA and the evolution of human-specific traits. *Nature* 471:216–219.

Milo, R., S. Itzkovitz, N. Kashtan, R. Levitt, S. Shen-Orr, I. Ayzenshtat, M. Sheffer, and U. Alon. 2004. Superfamilies of evolved and designed networks. *Science* 303:1538–1542.

Molina, N., and E. van Nimwegen. 2009. Scaling laws in functional genome content across prokaryotic clades and lifestyles. *Trends in Genetics* 25:243–247.

Ptashne, M. 2004. *A genetic switch: Phage lambda revisited*. Cold Spring Harbor, NY: Cold Spring Harbor Laboratory Press.

Shen-Orr, S. S., R. Milo, S. Mangan, and U. Alon. 2002. Network motifs in the transcriptional regulation network of *Escherichia coli*. *Nature Genetics* 31:64–68.

Somel, M., X. Liu, and P. Khaitovich. 2013. Human brain evolution: Transcripts, metabolites and their regulators. *Nature Reviews Neuroscience* 14:112–127.

8. Theft, Imitation, and the Roots of Innovation

Brändén, C.-I., and J. Tooze. 2009. *Introduction to protein structure*. New York: Garland Science.

Carroll, S. B. 2006. *The making of the fittest: DNA and the ultimate forensic record of evolution*. New York: W. W. Norton.

Deschamps, J. 2008. Tailored *Hox* gene transcription and the making of the thumb. *Genes & Development* 22:293–296.

Gilad, Y., O. Man, S. Paabo, and D. Lancet. 2003. Human specific loss of olfactory receptor genes. *Proceedings of the National Academy of Sciences of the USA* 100:3324–3327.

Glusman, G., I. Yanai, I. Rubin, and D. Lancet. 2001. The complete human olfactory subgenome. *Genome Research* 11:685–702.

Kirschner, M., and J. Gerhart. 2005. *The plausibility of life: Resolving Darwin's dilemma.* New Haven, CT: Yale University Press.

Knight, R., and B. Buhler. 2015. *Follow your gut: The enormous impact of tiny microbes.* New York: Simon & Schuster.

Ohno, S. 1970. *Evolution by gene duplication.* Berlin: Springer-Verlag.

Pal, C., B. Papp, and M. J. Lercher. 2005. Adaptive evolution of bacterial metabolic networks by horizontal gene transfer. *Nature Genetics* 37: 1372–1375.

Popa, O., E. Hazkani-Covo, G. Landan, W. Martin, and T. Dagan. 2011. Directed networks reveal genomic barriers and DNA repair bypasses to lateral gene transfer among prokaryotes. *Genome Research* 21:599–609.

Quignon, P., M. Giraud, M. Rimbault, P. Lavigne, S. Tacher, E. Morin, E. Retout, A. S. Valin, K. Lindblad-Toh, J. Nicolas, et al. 2005. The dog and rat olfactory receptor repertoires. *Genome Biology* 6:R83.

Schechter, A. N. 2008. Hemoglobin research and the origins of molecular medicine. *Blood* 112:3927–3938.

9. A Secret Life in the Shadows

Bodeman, J. 2003. Act fifteen. Mister Prediction, interview in "20 Acts in 60 Minutes," show 241 on *This American Life*, air date July 11, 2003, National Public Radio.

Ciccarelli, F. D., T. Doerks, C. von Mering, C. J. Creevey, B. Snel, and P. Bork. 2006. Toward automatic reconstruction of a highly resolved tree of life. *Science* 311:1283–1287.

Koonin, E. V. 2012. *The logic of chance: The nature and origin of biological evolution.* Upper Saddle River, NJ: Pearson Education.

Koonin, E. V., and M. Y. Galperin. 2003. *Sequence - evolution - function: Computational approaches in comparative genomics.* Boston: Kluwer Academic.

Lane, N., and W. Martin. 2010. The energetics of genome complexity. *Nature* 467:929–934.

Margulis, L., and D. Sagan. 2002. *Acquiring genomes: A theory of the origins of species.* New York: Basic Books.

Martin, W., and E. V. Koonin. 2006. Introns and the origin of nucleus-cytosol compartmentalization. *Nature* 440:41–45.

Martin, W., and M. Mentel. 2010. The origin of mitochondria. *Nature Education* 3:58.

Timmis, J. N., M. A. Ayliffe, C. Y. Huang, and W. Martin. 2004. Endosymbiotic gene transfer: Organelle genomes forge eukaryotic chromosomes. *Nature Reviews Genetics* 5:123–135.

van der Giezen, M., and J. Tovar. 2005. Degenerate mitochondria. *EMBO Reports* 6:525–530.

Woese, C. R., and G. E. Fox. 1977. Phylogenetic structure of the prokaryotic domain: The primary kingdoms. *Proceedings of the National Academy of Sciences of the USA* 74:5088–5090.

10. Life's Unwinnable War against Freeloaders

Doolittle, W. F., and C. Sapienza. 1980. Selfish genes, the phenotype paradigm and genome evolution. *Nature* 284:601–603.

Gould, S. J., and R. C. Lewontin. 1979. The spandrels of San Marco and the Panglossian paradigm: A critique of the adaptationist programme. *Proceedings of the Royal Society of London B* 205:581–598.

Gould, S. J., and E. S. Vrba. 1982. Exaptation; a missing term in the science of form. *Paleobiology* 8:4–15.

Gregory, T. R. 2005. *The evolution of the genome.* Burlington, MA: Elsevier Academic.

Kovalskaya, N., and R. W. Hammond. 2014. Molecular biology of viroid-host interactions and disease control strategies. *Plant Science* 228:48–60.

Martin, W. F., J. Baross, D. Kelley, and M. J. Russel. 2008. Hydrothermal vents and the origin of life. *Nature Reviews: Microbiology* 6:805–814.

Martin, W. F., F. L. Sousa, and N. Lane. 2014. Energy at life's origin. *Science* 344:1092–1093.

Orgel, L. E., and F. H. C. Crick. 1980. Selfish DNA—the ultimate parasite. *Nature* 284:604–607.

Wochner, A., J. Attwater, A. Coulson, and P. Holliger. 2011. Ribozyme-catalyzed transcription of an active ribozyme. *Science* 332:209–212.

Epilogue

Delbrück, M. 1949. A physicist looks at biology. *Transactions of the Connecticut Academy of Arts and Sciences* 38:173–190.

Acknowledgments

A long time ago, Doron Lancet suggested that teaching a course at the Weizmann Institute of Science in Rehovot, Israel, would help in the writing of this book. Ten years later (we had careers to attend to), we taught a revamped version of the course for students of biology, computer science, and the humanities, simultaneously at the Technion—Israel Institute of Technology in Haifa—and Heinrich Heine University in Düsseldorf. We warmly thank our many students for their enthusiasm and feedback.

We are grateful to our wonderful agent, Max Brockman, who patiently guided us through the adventure of publishing this book.

A lot of what we know about genes and their interactions is derived from enlightening discussions we have had over the years with our colleagues. In particular, we mention Peer Bork, Antoine Danchin, Charles DeLisi, Brian Hall, Tamar Hashimshony, Craig Hunter, Laurence Hurst, Ran Kafri, Marc Kirschner, Roy Kishony, Eugene Koonin, Doron Lancet, Eric Lander, Michael Levitt, Bill Martin, Ron Milo, Csaba Pál, Balázs

Papp, Leon Peshkin, Yitzhak Pilpel, Benjamin Podbilewicz, Aviv Regev, Daniel Segrè, and Zhiping Weng.

For reading and commenting on earlier versions of the book, we thank Gal Avital, Michal Gilon-Yanai, Vlad Grishkevich, Klaus Hartmann, Grün Kissenmann, Nina Knipprath, Claudia Larsson, Veronica Maurino, Asher Moshe, Avital Polsky, Joseph Ryan, Antonio Rodriguez, Leona Samson, Alex Shalek, Ori Spiegelman, Florian Wagner, Achim Wambach, Pamela Weintraub, Moshe and Rachel Yanai, and many others of our friends and students who pointed out ways to improve *The Society of Genes*.

We thank Bettina, Klaus, and Bruno Hartmann for providing the most beautiful quiet place in the hills of Penedo for focused writing. We also thank the Radcliffe Institute for Advanced Study at Harvard University for a wonderful environment for the editing stage.

Steven Lee made the amazing illustrations for the book, showing tremendous patience with our repeated requests for changes. We also thank Tamar Hashimshony for providing early versions of some of the illustrations.

Susan Jean Miller did a great job editing the final drafts of the book. We also gratefully thank our editor, Thomas LeBien, and Michael Fisher and Lauren Esdaile at Harvard University Press for their support.

Most importantly, we thank our loving partners and families for their enduring support throughout this adventure.

Index

Note: Page numbers in *italics* indicate figures.